MY PHYSICAL AND SPIRITUAL *Journey* INTO TRUTH

MY PHYSICAL AND SPIRITUAL Journey INTO TRUTH

My Journey into Truth

AINSLEY CHALMERS

LitPrime Solutions
21250 Hawthorne Blvd
Suite 500, Torrance, CA 90503
www.litprime.com
Phone: 1 (209) 788-3500

© 2021 Ainsley Chalmers. All rights reserved.

No part of this book may be reproduced, stored in a retrieval system, or transmitted by any means without the written permission of the author.

Published by LitPrime Solutions   06/12/2021

ISBN: 978-1-954886-88-9(sc)
ISBN: 978-1-954886-89-6(e)

Library of Congress Control Number: 2021912913

Any people depicted in stock imagery provided by iStock are models, and such images are being used for illustrative purposes only.

Certain stock imagery © iStock.

Because of the dynamic nature of the Internet, any web addresses or links contained in this book may have changed since publication and may no longer be valid. The views expressed in this work are solely those of the author and do not necessarily reflect the views of the publisher, and the publisher hereby disclaims any responsibility for them.

Contents

Introduction ix
Background of Family History 1
Early Days in India. 3
Early Days in Australia. 5
Early School Days in Australia. 8
Post School Days 10
Postgraduate Days 15
Tenured Positions Back in Adelaide 18
My Departure from God 20
My Return to God 22
My Life Post Conversion 25
Life at the Mater Hospital in Brisbane, Queensland 27
Research at the Mater Hospital 29
Return to Flinders Medical Centre (FMC) 34
Our Stress Work. 40
Xanthochromia Work. 44
My European Trip with Krista Marcher. 46
Return to Adelaide in 2001 after the European Holiday 49
Spiritual Aspects of My Journey in Life 52
Spiritual Lessons Learnt. 54

There Are Two Basic Sciences – Experimental and Historical..78
Naturalism versus Biblical Creationism 80
Geologic Prodding 83
DNA and Evolution 86
Evolution and Cosmology.............................90
Bad Press on Religion.................................93
Conclusion 95
Publications – A H Chalmers........................99
Abstracts of Scientific Meetings – A H Chalmers 107

This book is dedicated to my Lord and Saviour Jesus Christ who as you will see has led me with much love and grace throughout my whole life.

There have been many who have been used by the Lord to lead me in my growth as a Christian and I thank God for them. Included in these is my wife Denise (dec 1991), my daughters Joanne, Nicole and Jacqueline, Bob McGregor (dec), Rudi Roll, Leonie Lindsay, my grandkids Joshua, Matthew, Isaac, George, Charlie, Rubi Rose and Lucas.

I have also been greatly helped in my Christian walk by Creation Ministries International, GOT ministries (e-bible) and Our Daily Bread studies. There are also many others too numerous to mention and some are mentioned in this book. I apologise for omitting theirs and others names on this page but God knows who they are.

Finally I would like to acknowledge grants from these bodies that permitted much of the research described in this book. They include: Adelaide and Flinders University Research grants, South Australian State health, The John P Kelly Foundation at the Mater Hospital Brisbane, Australian Cancer Research grants and finally the National Health and Medical Research Council of Australia.

Introduction

I felt prompted to write this book towards the end of my life so that my family will have some knowledge about it. It has occurred to me that family histories go down to the grave unless they are documented. In that sense I wished that my parents had written something about their lives so that I would have had a greater knowledge and understanding of my history.

Though I came to the Lord Jesus Christ over forty-one years ago, I have felt, looking back in retirement, His leading me in my life, including the research areas I have been involved in and discussed in this book. At the time I had a feint inkling that God was leading me, but in looking back, I can clearly see His hand in my life. And so this book attempts to describe His supernatural leading in all aspects of my life including into the truth of His words, the Bible. I have called this leading *spiritual experiences* since they display the invisible hand of God. In short, this book is not meant to be a big ego thing, just an autobiography hopefully illustrating both physical and spiritual aspects of my life's walk into truth.

Background of Family History

My sister-in-law, Maureen Chalmers, competently searched our family history and all thanks are due to her endeavours in this area. My dad's side dates back to 1846 when Robert Fleming Contes Chalmers was born in Glasgow. Sometime later he was in a Scottish regiment that was posted to India where he died later in 1886. He married Julia Rosamond Winefred Boyle and later my grandfather Mark Howard Chalmers was born in1886. Mark married Gentle Rees Isitt, and they had two boys Roland Fleming Chalmers and my dad Richard Arnold (b 1915). My dad married my mum, Olivia Theresa McKenzie, in 1937, and the five children resulting from this union are described below (see 'Early Days in India'). Mum's family dates back to Adam Gordon (b 1783) in India, who married Louise Lally. They had several children one of whom became my grandmother namely Gertrude Eveline Muriel Gordon (b 1896). She married Lionel Weston McKenzie (b 1889), and they had eight children with my mother Olivia Theresa McKenzie being born in1916. To cut a long story short the Chalmers time in India date back to the mid-1800s and the Gordons to the mid-1700s or perhaps even earlier.

A comment needs to be made about my grandfather Mark's life. Mark Chalmers's first wife Gentle, a Jewish lady, committed suicide a few years after they divorced, in about 1923. Mark became a prolific walker hiking 300–400 miles at a time through the Himalayas, traversing Tibet and Nepal on a few occasions in the 1930s, maybe in an attempt to overcome his grief. He documented his walks. Unfortunately I was unable to find all the publications of his treks, particularly his walk from India to England in 1933. His arrival in London in August 1933 was first page news in the *Daily Express* newspaper. I don't ever recall ever meeting any of my grandparents.

Mark Chalmers later married another woman called Louisa who apparently was not very nice coming between my grandfather and his two boys, one being my dad. No more on that gossip.

Early Days in India

My life started on 5 August 1938 in an Indian town called Lucknow. My family spent a few years there then moved to a town on the foothills of the Himalayan mountains called Naini Tal about 7,500 feet above sea level and about 400 km north of New Delhi. Before moving there my brother Geoffrey was born on 28 December 1941 in Lucknow.

Naini was a beautiful place very much like a Swiss village with a long lake in the middle of the town, about one mile long by about half mile wide. In winter the town was covered in snow, and as a young boy, I remember climbing the snow onto our rooftop. The population of Naini in those days was probably about 20,000, a small town by Indian standards. Why we settled there I have no idea except that it was a cool place during the unbearably hot Indian summers. Also my dad and his brother Roland had prior association with Naini being educated at St Joseph's Catholic College.

The other interesting aspect of Naini is that it was in the Kumauni region, an area well known for its wildlife especially tigers, leopards, hyenas, cobras and others. The tigers were dangerous and killed Indian labourers working on a nearby rail line by removing them from their tents at night. To protect the Indian workers, a local army person called Major James Corbett had the job of killing many tigers.

The species are now protected and anyone killing them today would get a long jail term. Nonetheless the area, now a national park where James Corbett did his culling work, is named after him.

The other aspect of Naini I remember is that the leopards used to visit our homes at night and eat our pet dogs. However, a friend's dog, a bull terrier, actually killed a leopard although the dog died a week later from its wounds. The other aspect that I didn't like about Naini is that the hyenas used to keep us awake at night with their weird laughter.

I remember one night hearing a scream from the cook in the kitchen. Dad and I ran in to investigate only to find the cook perched on top of the kitchen cupboards and a cobra snake in attack mode was looking at him from the floor. Dad immediately grabbed a broomstick and broke the cobra's back.

In Naini I did my first three years of schooling at Dad's old school, St Josephs, run by Irish Christian brothers. I don't remember a great deal about it but later returned to it as an adult (see later chapter).

We then left Naini in 1948 except there were now three boys with Richard being born on 2 September 1946 in Naini. We gradually travelled south staying short term in Meerut, Delhi and Deolali. We decided to leave India because, after independence from the UK in 1947, the country became unsafe for white folk living there. Dad tried to get us to Australia, but we initially failed because of the White Australia policy of those days. Not that we were black but any hint of colour could stop one's migration to Australia. How things have changed for the better nowadays. Anyway one family pulled out of their Australia trip, and we took their place. We were on the verge of migrating to the UK but instead went by ship the *Empire Brent* to Australia in April 1948.

Early Days in Australia

We arrived in Melbourne on 15 May 1948 and rented a flat in St Kilda, a seaside suburb of Melbourne. After about two months, we headed to Adelaide, South Australia, because friends from India the Huttons had settled there. We lived initially for about three months in their single garage at Hampstead Gardens, a suburb twelve kilometres north of Adelaide. Dad then bought a house at Brighton, a beachside suburb fifteen kilometres south of Adelaide. It was basic and half completed with no internal walls or ceiling but had six blocks of land around it. Dad being ever industrious put up the walls with Mum and my limited help, then a 10-year-old. Later Dad built two sleep outs in the front of the house, and those were to be Geoff's and my bedrooms.

The Brighton house was located at 83 Gladstone Road, Brighton, only 100 metres from the beach. Although it wasn't popular to live by the beach in those days, we had a natural playground at our disposal. The house was surrounded by huge sandhills and us brothers lived in the sea even though we couldn't swim. We had a neighbour's cows in the backyard and an outside toilet about ten metres from the back door. To go to the toilet, you had to first brush away the redback spiders. Huge rats also visited us at night as they loved to feed on our hair. Dad used to go to bed with a baseball bat, and during the night, one would hear huge whacks as he hit home runs with the rats as balls. Sewerage tanks served the house, and Dad and I would have to service them every few months, a very smelly job.

Living in Australia in the 1940s was pioneering stuff and hard work for Mum and Dad as they had servants in India. There were no dishwashers, automatic washing machines, and hot water on tap. Mum had to learn to cook and Dad had to be a handyman as there weren't many builders around in those days. Dad did the electrical work in the house, and it wasn't quite right as we used to get a shock when opening the fridge door, so we wisely opened it holding a tea towel. We owned a 250 cc BSA motorcycle and sidecar, and with all five of us in that contraption, it was a sight to behold. There were no supermarkets, only small street-corner grocery shops. In our 1 km long street, there were only two other families, the Rances and the McArdles. Davy Crockett would have felt very much at home there.

Before we got a fridge, we had an icebox with the iceman coming up the street in his horse and cart selling ice blocks every second day. We also had horse carts delivering milk and bread.

We had no showers or hot water on tap. Dad heated water in a large copper container (called a copper) with wood as fuel, and once a week we would have a bath. Clothes washing also needed hot water, and Mum would stir the washing in the copper and then, after cold water rinsing, put the clothes through a manual arm-operated ringer to squeeze the water out. With the big bed sheets, Mum would grab one end and I the other, and we would twist it in opposite directions to remove most of the water.

Early on we had no TV (it came to Adelaide in 1959). Our main entertainment was an old Mullard radio, and about one mile away, the Windsor movie theatre. It still operates with the same old time atmosphere inside. In those days, movies were mainly black and white and corny by today's standards, but to us it was a magical place of escape from the tough times our parents endured. The other entertainment was for Mum, Dad, and me (a then 12-year-old) to play poker with my aunt Clare and her husband Denzil. We gambled with 1–6 penny bets per game. The games usually lasted all Saturday night till early Sunday morning. In those days nearly everyone smoked and so I would play poker in a fog of smoke. It was a miracle that I survived all that passive smoking. Anyway it all served a purpose in that I grew up hating smoking and gambling.

My parents had two more boys born in Australia; Mark was born on 10 May 1953 and Michael, on 3 October 1954. Financially it was tough times. Dad initially worked at General Motor Holden on the car assembly line but then later got a clerical job with birth, deaths and marriages, a position more suited to his clerical abilities. He eventually rose to be deputy registrar and performed many marriages as a celebrant. To help feed five hungry mouths, Mum ran a crèche with the help of a young girl. Sometimes fifty children were at the crèche. She did this onerous task for about three and a half years before the council stopped the operation of crèches as the area became more up market. Gradually Dad sold the extra blocks of land around our place to help educate, clothe, and feed us boys.

Before her marriage to Denzil, Clare, Mum's youngest sister lived with us at Brighton. She, being six years my senior, was like a teenage sister to us boys and worked at Minda Home, walking distance from our home. Minda was a home for people with mental disabilities such as Down's and other similar conditions. There were no fences around Minda and so these lovely folk would sometimes walk casually into our house asking for cigarettes from our parents. In return they would give us reams of comic books.

Our home life had a lot of drama in it in that Mum and Dad fought a lot due to stresses involved in raising a family in those days on limited resources. I sometimes took my younger brothers to a nearby sandhill so that they wouldn't hear the yelling and screaming. Nonetheless my parents did a fine job in bringing us up, and I have nothing but love and admiration for them. They were wonderful ballroom dancers, and as a kid I loved to see them dancing in our small living room. They always won first prize for dancing at the annual church ball. I still get tearful thinking about the seamless yet loving way that they danced together. I feel sure that Fred Astaire and Ginger Rogers would have been amazed by their dancing ability.

Early School Days in Australia

My primary schooling was at St Theresa's convent Brighton, about one mile from home. Sisters of Mercy were our teachers, and they made fun of my speaking in an Indian accent complete with wobbly head. For example, they would say things like 'Where are your bow and arrows Indian?' and 'Did you come to school on your elephant Indian?' Daily they incessantly beat me verbally and physically because of my being so different from the *normal* Australian. There wasn't much mercy in those nuns! Naturally, the other boys in my school also bullied me all the time. After a few months of this treatment, I learnt to fight back with some coaching in boxing from Dad and discovered that I had a talent in boxing. To this day I have a deep hatred toward bullying of any sort.

Because of being put down so much, my self-esteem was very low and I used to come last in class. Despite this cruelty, I went to mass every Sunday at this school. Mass was a jumble of Latin words, but there was something spiritual there that drew me to go every Sunday. The McArdles living across the road from us took me to mass in the Dickey seat of their T-model Ford.

After completing primary schooling (grades 1–7), I went to Sacred Heart College (SHC, grades 8–12) where Marist brothers taught us. The brothers were no better than the nuns when it came to cruel behaviour towards us boys. They verbally put us down and were also physically cruel, caning us for minor infringements. They very possibly were sexually frustrated, having taken the vow of celibacy.

At SHC the brothers pushed us into playing sport, and I ended up playing Australian rules football in winter and athletics in summer. On top of this I was involved in the school army cadets. In grade 8, I used to come in the bottom 5 in a class of about 50 students. In grade 9 my grades weren't any better; however, halfway through the year a significant event happened to me.

An older boy from a higher grade started to bully me, and so one day I retaliated with my fists and beat him up. Brother Damian, the head principal, wasn't happy about this fight. I was caned on the hands and ordered to stay back after school and write out 100 times, 'I must not fight in school.' To save time, I wrote my lines out with bleeding fingers during the Latin lesson taught by Brother Damian. Well, he caught me writing my lines and told me that I was no longer welcome at the school and to take my books and never come back. I was expelled.

I went home and didn't tell Dad what had happened because he would have beaten me up as well. So the next day I slinked quietly back to school and kept a very low profile. That next term I tried to be the model student doing my homework conscientiously every night. At the end of term, we had our exams as usual and much to my surprise I went from the bottom five in class to the top five. The brothers were as shocked as I was, but it made me realize that maybe I wasn't as dumb as I thought I might be. To cut a long story short, at the end of year ten, I was standing at the Adelaide town hall receiving prizes (usually books) for academic excellence from the then governor of South Australia, Sir Willoughby Norrie. What a turnaround!

In year 11 I continued to do well academically, especially in the science subjects of maths, physics and chemistry. At the end of year exams, I was awarded a Commonwealth scholarship because I came in the first 250 in the state of South Australia. This award would pay my university fees except I had no intention or thought of doing tertiary education. I hadn't even planned for it.

Post School Days

With a little research, I found that my grades in high school would have permitted me to do a medical or science degree. Even with a scholarship, Dad said that he couldn't afford for me to do either of these degrees due to family expenses. So Dad, in his wisdom, thought I should do economics and accountancy at University of Adelaide whilst working in the Commonwealth bank. So at 17 years of age instead of doing year 12 at school to prepare for university, I followed my dad's leading, as well intentioned as it was. Well I hated the bank job and the studies but persisted for the year. I failed my university courses miserably. The only thing that kept me going that year was the 250 cc BSA motorbike I bought with my bank earnings. And that leads onto another story.

After buying the motorbike and owning it for only about two days, I rode it to university to do an economics subject after work. I couldn't wait to ride my shiny bike home after the lecture. I offered to take a male student friend home on the back of my bike as he lived near me. When I tried to kick start it, the bike backfired through the carburettor and immediately went up in flames. Cars near my bike drove off frantically in all directions, and my bike very nearly set fire to the university library called Barr Smith. The fire reached a height of about three metres and was lapping the electric wires overhead. Well, we both caught the train home, leaving one very charred bike at the university and one very grief-stricken boy in tears, the joy of my life completely cremated.

The next day was a Saturday and Dad and I collected the bike and took it home. We both worked on that bike for the next month and slowly brought it back to life. It never worked well, and I used to push it more than ride it. After 3 months of heavy pushing, I sold it and bought a 120 cc Honda two-stroke motorbike. It was a great bike and never let me down once.

My National Service Experiences

When I reached my eighteenth birthday I haad to do compulsory National Service. This comprised 77 days of full time service at Woodside Army Camp located in the Adelaide hills about 30 kilometres from my home in Brighton. This initial service was followed up by 2-3 week bivouacs over the next 3 years.

At Woodside we were trained by full time army personnel some of who were very cruel. We were yelled and screamed at all the time and made to do cruel things like shovel dirt into a hot head wind. We were let out for weekend leave which was like a great escape with all of us heading furiously for home on our motorbikes and cars. If you were naughty you lost your weekend leave and spent the whole weekend pealing potatoes or cleaning the camp. It was considered a fate worse than death. In camp we learnt to parade, clean and fire our 303 rifles, Bren guns, Vickers machine guns and throw grenades. We also had to complete certain physical feats like running in full army gear for 5 miles in less than one hour, climb rope ladders, crawl on your stomachs through mud with rifles on hand. We also had to complete obstacle courses which were really hard work. We lived in long tin sheds with about 20-30 folk in each one and every morning your bed had to be made to an exacting standard before breakfast. The army really taught discipline.

Despite the hard times there were some fun times as well. In our shed were had one prtson who was an expert hypnotist and hypnotised some of the guys. At one time he hypnotised two people into believing a third person, lying innocently in bed and reading a book, to be Marilyn Monroe. Well these two tried so hard to woo "Marilyn" who was fighting them both off. On another occasion he got one person to believe that they were a particular nasty sergeant. Well this "nasty one" were ordering the whole platoon onto the parade ground at 10 pm at night and made us parade bear footed. He really sounded like the real nasty sergeant. Most in the platoon couldn't be hypnotised and I was one of them.

After the initial training I was seconded to the engineers for the 2-3 week bivouac training period over the next 3 years. As engineers we were taught explosives, clearing mine fields and bridge building. One year we went to Sydney sitting upright for the 3 day train journey. In Sydney we saw TV for the first time (1956), the Harbour Bridge and Luna park. It was all very exciting except for one night. We were woken up at midnight and told we had to delouse a minefield alongside the Georges river. The minefield contained mines however with less explosives, only enough to blow off your foot. To locate the mines you had to gently prod the soil with a bayonet. A solid object inder the soil indicated a mine the location of which you then marked with a red flag. While this process was going on tracer bullets were being fired over our sweaty heads not to mention the mosquitoes feasting on us. We eventually blew up the red flagged mines after 2 hours of bayonet prodding. Some of the servicemen understandably broke down crying during this stressful exercise. We also built Bailey bridges across the river as well.

The following year we camped in the desert about 200 kilometers north of Port Augusta, a place called El Alamein. All we had for company here was salt bush, red back spiders (poisonous) and scorpions. The bush was one's toilet and because of the nasty insects, we slept the whole ten days in our army uniform, boots and all. Nonetheless some soldiers were carted of to hospital with bites from these insects. The temperature was over 40C everyday and we had to build bridges over ravines in this heat. Some of the soldiers collapsed on the ground frothing at the mouth. They were diagnosed with salt deficiency and so every morning we all had to take a large salt tablet.

One day we had to transport 4 x 40 gallon drums of petrol about 5 miles across the gulf. To do so we used two army amphibian ducks. After dropping off the petrol, a storm sprung up on the way back and one of the duck's engine stopped working. The remaining working duck had to then put a tow rope to the non-functional boat and tow it home. Attaching a tow rope in a violent storm was not easy particularly as most of the crew were hopelessly seasick. The remaining few of us eventually managed to do so but it was akin to doing so on the back of a wild bucking bull. After we arrived soaked on the shore we were then driven in an open truck arriving at our camp at three am covered in red desert dust. We were on duty at six am the next morning feeling very tired and dirty.

When Natuinal service ended it was a great relief although one did miss the camaraderie. We were lucky in that we were too young for the Korean war and too old for the Vietnam war. National service was hard work but no where as bad as being in a real life war situation.

The start of my science career

At the end of my one-year stint at the bank, I quit and found a job as a junior laboratory assistant in the SA Department of Mines doing elemental rock analyses. At night, I did my bachelor of science degree studies. It was tiring doing both, namely work and study, but I loved science and I loved my job. Besides, I was young and had loads of energy.

During my studies, I sold my bike and bought an FJ Holden sedan car, as I needed more room to carry all my books. At the weekends I used to go dancing at the Wonderland. It was ballroom dancing in those days and a chance to meet eligible young girls.

At 20, I met a lovely girl Libby. We became engaged, but religion was a stumbling block to our relationship. I was Catholic and she

Methodist. We were going to compromise and join the Anglican Church but my Catholic priest told me that I would go to hell if I did so. Well, the stress of going to hell broke up the relationship, very sad. In those days of the Latin mass, if the church said jump, you'd say how high. In the1950s denominations were important. Nowadays it doesn't matter a hoot. Libby invited me to go to a Billy Graham crusade meet at the Adelaide oval in February 1959. My Catholic upbringing put a stop to that. In hindsight, this is one of the major regrets of my life that I didn't go to that meet. After splitting, Libby went on to meet a nice man and married and had four children. Well I passed my science subject, some with credits and distinctions and in the final year got a scholarship to complete that year. There were problems at home (it was noisy) and so I rented a single-bedroom close to the University of Adelaide. It was a very lonely year, especially so soon after losing Libby. At the end of the year I got two high-distinction passes for my final year in chemistry subjects.

After my BSc degree I met my future wife Denise at the Wonderland, and after two years, we got engaged and then married at SHC chapel on 11 July 1964. We were both Catholics, so there were no religion problems.

Postgraduate Days

My chemistry professor wanted me to do an honours research degree, but my SA state scholarship wouldn't allow it for at least two years. In those two years I worked at the SA State Department of Chemistry in Kintore Avenue, Adelaide, doing mainly forensic analyses. The job was good but not challenging enough, and after the two years were up, I took up a job as a scientific officer in the Commonwealth Scientific Industrial Research Organization (CSIRO) Department of Chemistry at Fishermen's Bend in suburban Melbourne, Victoria. At this job I isolated and helped identify alkaloids from plant material, which were toxic to animals. It was a good job but not quite where I felt that I should be, wherever that was. After Denise and I married on 11 July 1964, she came to live with me in Melbourne in a suburb called Murrumbeena where we rented a simple one-bedroom flat at the rear of a home. Denise worked as a secretary with General Motors Holden and later with Mauri Brothers and Thompson, a food distributor. Denise became pregnant, and we both decided to return to Adelaide as both our families lived there. On 5 December 1965 our beautiful daughter Joanne Michelle was born at Calvary Hospital in North Adelaide. I was literally walking on air being so joyful.

In Adelaide I got a job in Australian Mineral Development

Laboratories doing elemental analyses once again. The salary was good, but I felt that I was called to do something else in my life. I prayed to God and eventually found an opening for a professional officer in the University of Adelaide. My salary dropped by half, but I felt that I was in the right place, for once! My job was to set up a biochemistry lab in the department of surgery at the Royal Adelaide Hospital. The job was challenging but I loved helping clinicians with their research work.

After hours, since I didn't have an honours degree qualification, I did a master of science qualifying exam at the new Flinders University of South Australia (FUSA). For my research component of my master's degree, I worked with Maurice Atkinson, professor of biochemistry at FUSA and Dr Peter Knight, head transplant surgeon with the department of surgery. We worked on the mechanism of action of a thiopurine drug (azathioprine), which allowed success in the renal transplant field.

Disaster struck in the third year of my studies when Maurice died prematurely at the age of about 35. Andy Murray replaced Maurice and became my supervisor. I was greatly blessed by having both these talented scientists as my supervisor. Because the work went so well with about 12 peer-reviewed publications, I was awarded a PhD.

In 1966, we moved into our first home. In March 1967 we lost our second child named John, who died on his birthday, a stillborn baby. It truly was a very sad moment especially for Denise who carried this baby to term and suffered greatly with grief. Nicky was born on 30 June 1968 and joy once again flooded back into our lives.

My PhD studies showed that lymphoid tissues rely on DNA precursors (called purines) made in the liver. These were then transported to lymphoid tissues by blood. The thiopurine drugs were inhibiting both liver syntheses and transport of purines thereby limiting DNA synthesis in the lymphoid tissue resulting in immunosuppression and the transplant not to be rejected. In 1972 I was awarded a PhD by Flinders University.

In 1971, I did a cancer postdoctoral fellowship at the University of Queensland with Professor Chev Kidson. He was a very stimulating man, spending much of his research life at Stanford University in America. We worked on human malignant melanoma cells that we grew in culture. We measured the effect of UV on mutational changes and repair in these cells and had a productive research time.

The main research finding was that melanoma cells were resistant to UV irradiation compared to normal healthy cell lines, leading to the conclusion that there may have been a mutation in the tumour suppressor gene (SG). In normal circumstances the SG slows down DNA synthesis so that UV inflicted damage to the DNA may be repaired. We reasoned, therefore, that if repair is not carried out, then this can lead to further specific deleterious mutation events, which could enhance cancer growth and metastases.

On 22 May 1972 Jacqueline our third was born in Brisbane, another joyful event. Much as I loved the cancer work, I was on soft research money. In short it was untenured and my salary could dry up very quickly. I could find myself unemployed with a wife and three daughters to support.

Tenured Positions Back in Adelaide

At the end of 1973 we returned to Adelaide where I got a job as a medical scientist with the Institute of Medical and Veterinary Science (IMVS). This job was tenured, much to my relief, having five mouths to feed and a mortgage to meet. The job meant doing routine clinical enzymology as well as research arising within the department.

One research program we did was to try and work out why some patients in ICU on a parenteral sugar called xylitol were dying from oxalosis, that is calcium oxalate crystal deposition in various body tissues including the brain. The bottom line of this finding was that those patients with depleted B vitamins shunted xylitol metabolism into oxalate. As a result xylitol was removed as a parenteral nutrient, replaced by glucose, and the recommendation that further infusions required also the addition of vitamins to patients on parenteral nutrition. So this was a significant clinical finding with life-saving repercussions.

In 1975, the new Flinders Medical Centre (FMC) was opening up. This hospital offered medical degrees and was affiliated with the Flinders University of South Australia. It also had a significant commitment to research in the medical field. I transferred from the IMVS to FMC and initially one of my primary tasks was to set up part of the clinical biochemistry section called special chemistry. Getting these specialized assays going was a challenge, and we published some aspects of this.

I was also involved in teaching medical biochemistry to first- and second-year medical students. The medical course at FMC was the first postgraduate medical course in Australia and therefore allowed only graduate and postgraduate students into the medical degree. I was also involved, as part of a team, in selecting students deemed suitable for this course.

Just as an aside an amusing, though serious, incident occurred while I worked in the lab at FMC. The intensive care unit (ICU) of FMC was due to open in the 1970s. For those unfamiliar with the ICU section, it is an area set aside for the very sick patients, usually those unable to eat or breathe on their own. The opening was to be officiated by the then premier of South Australia, Don Dunstan and other VIP governmental and senior hospital officials. A few days before opening, the air and oxygen gas lines were tested to ensure that they worked up to standard. To the horror of the testing engineers involved, a foul smell emitted from these lines.

Being in charge of the special chemistry section, I was called in to investigate together with Professor Malcolm Thompson, head of the school of chemistry. Well I froze the emitting smell in dry ice and analysed it using a gas chromatograph. It turned out to be the hydrocarbon solvents (HCs) used for degreasing these gas lines before installation. The contractor used hydrocarbons to degrease the new lines instead of a bicarbonate solution usually used, thinking that the degreasing job would be simpler and cheaper because the HCs would evaporate off the washed lines. However, he failed to realize that some of the HCs had a high boiling point and were non-volatile. Consequently, there ensued a huge and expensive job making sure that those lines were properly cleaned. The ICU opening went ahead; however, the patients were administered bottled gases.

My Departure from God

In 1976, my spiritual life took a downturn. As a Catholic Christian, I started to doubt the reality of God. I kept searching for signs that God was real, but there appeared to be none. Working in a secular university environment with mostly atheistic scientists probably didn't help though I cannot blame this for my falling away from God. Toward the end of 1976, I had an affair with another married woman and left my lovely wife and three beautiful daughters. The lady involved also left her husband and two kids, so it was pretty messy to say the least. We rented a house about 15 kilometres from our respective spouses. On weekends we would take our kids, separately of course, to their activities that involved sport, horse riding and the like. The weekends were hell watching the pain of rejection on my wife's and children's faces.

My partner was very supportive of me, but the guilt of what I had done weighed heavily on me. To deaden my guilt, I started to drink wine. All in all, it wasn't a happy time, and I felt completely out of control and hopelessly lost. And it was all self-inflicted. About six months after leaving my wife Denise, she called me one day at work to say that she had met a man. Initially I thought a boyfriend and felt happy for her. It turned out that he was a happily married Christian pastor with a wife and kids. Anyway she wanted me to

MY PHYSICAL AND SPIRITUAL JOURNEY INTO TRUTH

meet him as he lived close to FMC, my place of work. So on 25 July 1977 I met Denise on my lunch break, and she introduced me to Pastor Bob McGregor in his lounge room. Bob was a minister at a nearby Pentecostal Church called Bethesda.

Bob was active in the deliverance ministry, and when asked, I gave him a rundown on my so-called spiritual life. When I told him I had been involved with Ouija board games, he suggested that we needed prayer to break the spiritual hold that this had had on my life. He also looked deep into my eyes and said that I was a tormented soul. At that time what he said didn't make much sense, except for the tormented soul bit. But he then asked me to accept Jesus Christ as my Lord and Saviour, a term then foreign to me. I went along with him not totally convinced but mainly out of pity for Denise. He led me through the sinner's prayer. Well I said the prayer and all was well till he very gently laid hands on my head and asked for an infilling of the Holy Spirit. When he did that, I felt a peace running through me that made me want to fall to the floor and start crying. But I held back with all of my strength. I didn't, at that time, understand fully what was happening. It was all a new experience, the spiritual significance of which I was fully unaware at that time.

Well I went home to my partner that night and related what happened. She was appalled that I had been so horribly treated. Anyway we joked about my *deliverance* and had a drink to celebrate or rather mock this event. So life went on that week as normal. The following weekend on Saturday night, we were invited to a wedding. It was a fun night, but in the middle of the fun, out of the blue, I turned to her and said that I wanted to be alone. She couldn't understand my behaviour and neither could I. I had a deep conviction that I had to be alone.

The following week my partner tried to get me to change my mind, but I was set on having a short time (one week) on my own. I had forgotten Bob's prayer and just couldn't work out what was happening. Anyway my partner eventually relented and went to spend one week with her favourite aunt.

My Return to God

The day my partner had left, Denise rang me, out of the blue, at work and asked me to have dinner with her and the kids. That night, at her invitation, I stayed with Denise. I had a peace being reunited with my family, the first time in six to eight months. Eventually word had got back to my partner that Denise and I had reunited, and I presumed then that she had returned to her husband who I knew would have had her back in a flash.

When I had returned, Denise shared how she had become involved in the Catholic charismatic movement and also went to the Bethesda Pentecostal Church where Bob ministered. I told her I was happy to hear of her deeper walk with God but really didn't want to be a part of it. One week later whilst reading the paper in the lounge, I suddenly fell to my knees in a state of great depression and felt that suicide was my only option in life. That coming weekend I thought of throwing myself under a bus and making it quick. Denise sensed my low mood and suggested that I talk to a Catholic charismatic friend she had met called Rudi.

So after a Chinese meal on Saturday night, we proceeded to Rudi and his wife Shirley's house. They were a friendly Hungarian couple, and we sat in the lounge viewing a romantic comedy on their TV. Denise turned to Rudi and asked if he would speak with me and so we left the TV and went into the kitchen. I had no idea what this large Hungarian man was going to say, but he started talking about the Holy Spirit. I had no idea what was going on but out of politeness sat and listened to him ramble on. But then a strange thing happened. Whilst he was talking, I had a vision during which time his voice became a distant sound.

Let me be clear, I am not one to have visions, but this was a life-changing one. In this vision I saw myself on a stage where God clearly showed me how Satan had tried to destroy me over many years. I didn't believe in the demonic realm, and if anyone had tried to convince me of it, I would have left the room faster than Usain Bolt. However, in this vision I was completely convinced and what had happened in my life made complete sense. I knew, at that very moment in the depths of my being, that Jesus Christ was the way, truth and life and that I would have to walk closely with Him otherwise my life would be short. I had an immediate God-given different perspective on life. When the vision faded, Rudi stopped in the middle of his Holy Spirit talk, looked at me and said in an enthusiastic voice that I had been baptized in the Holy Spirit. The only manifestation was that I had tears flowing down my face. They weren't sad tears but ones of great joy. So all this happened on about 3 August, two days before my thirty-ninth birthday, in the fortieth year of my life.

When I drove home that night, I was as quiet as a mouse, and Denise was wondering what had happened. I was literally floored by so much godly revelation as a huge peace descended on me. The next day I related to Denise what had happened in Rudi's kitchen. I also knew without a shadow of doubt that I would have to walk very closely with my God, Jesus Christ. The choice was simple, follow God or die a young age.

After the Holy Spirit baptism, the word of God, the Holy Bible, became so real to me. I couldn't stop reading it even when I went to the toilet or on a bus going to work. Before the Holy Spirit baptism, the Bible would put me to sleep in five minutes flat whenever I read it, usually once in a blue moon. But now it became so alive to me. I wasn't sure where God wanted me and so I returned to my Catholic roots except this time I became more involved in the Catholic charismatic renewal. I enrolled in a correspondence course (a diploma in Christian studies) run by a group in Sydney.

Whilst Mum and Dad were on holiday in West Australia, the Holy Spirit had touched my mum one month before me. Dad, in retirement, had driven in a Kombi van around Australia with Mum. At 300 kilometres north of Geraldton, Dad had problems with the motor and in his impatience, had thrown petrol on the hot motor. The flashback set fire to his arm, and because Mum couldn't drive, he had to drive the 300 km to Geraldton with his arm cooling outside the van window. When they got to the hospital, Dad was immediately admitted and spent the next 1–2 months there as a patient.

Poor Mum was on her own in the van at a local caravan park in a remote town, not knowing anyone. Eventually in town one day she was approached by some Pentecostal women and invited to a local Assembly of God (AOG) church. Mum was always a good Christian woman. Nonetheless, at this country church she re-received Christ into her life and then received the Holy Spirit baptism one month before I did in Adelaide, 3,000 km away. Of course I had no idea of what was going on. Mum and Dad had lost contact with me over the period of my infidelity, but they did support Denise and my girls as much as possible. And I am eternally grateful to God and to them for that support of my family.

My Life Post Conversion

Toward the end of 1978 we all—Mum, Dad, Denise, my girls and I—started to go to a new Assembly of God (AOG) church near where we lived. My aunty Clare, Mum's youngest sister, primarily initiated this move. Well the minister took me under his wing and got me involved in many church activities like taking Bible classes, mini-preaching assignments, door knocking, etc. It was all a bit much for a pew-warming Catholic; nonetheless, I quite enjoyed my new Christian roles that greatly helped me to grow spiritually.

As I grew spiritually, Denise understandably got more and more resentful due to the fact I had damaged the family so much and here I was being hailed as a hero within the church. To make matters worse, I took up musical comedy singing on stage and was given, from the start, principal roles due to my semi-operatic bass baritone voice. In these roles I was usually the hero singing to my heroine, which didn't go down well, as one would imagine. Also the applause and adulation from the audience didn't help. So after one year, I gave up my musical career to help keep the peace at home. To be honest there was also considerable stress in attempting to play two roles, one as a senior hospital scientist the other as an entertainer.

As Denise became more and more resentful of me, I backed off some of my lead roles in the church in order to keep the peace. I was reaping what I had sowed. Despite knowing that Christ had forgiven me, I carried a huge guilty conscience with me wherever I went. And Denise because of her hurt kept this guilt very much alive. At FMC I had a very disagreeable boss with a bipolar disorder who had it in for others and myself within the department. To add to all my woes, Jackie my youngest daughter had very bad asthma, which required about three stays in hospital every year. I prayed earnestly to God about my situation and then in mid-1982 I was offered the inaugural laboratory manager position at the Mater Hospital in Brisbane.

Life at the Mater Hospital in Brisbane, Queensland

I accepted this job despite lesser pay with the hope that God in Brisbane may somehow help heal my marriage. Also and more importantly, it was hoped that Brisbane, being in the warmer tropics, might help improve Jackie's asthma condition. Denise also thought that this job was God leading us, and we were in His will. To be honest I never really discerned God's will, but trusted Denise in this matter. So in August 1982, I started work at the Mater Hospital Pathology Labs in Brisbane. Initially the job was a bit of a disappointment. It was all admin that I hated and zero research. I really missed doing research and cried and wept privately to God about it. There were also huge stresses in the family with Nicky missing her old school and hating her new one. I used to spend many hours playing tennis with Nicky to help let out her frustrations. In addition, Jackie's asthma got worse in the mouldy tropical weather. Also there was an underlying tension between Denise and me that the tropics didn't heal. So, all in all, life wasn't a bowl of cherries.

On the positive side, I had a great boss who valued my opinions about the running of the pathology area. The staff, that numbered about ninety, was also very friendly and supportive. But research was basically very low key, almost non-existent. There was a small library associated with the hospital with very limited number of medical publications when compared with FMC.

We bought a house in the suburb Wishart and initially started going to a nearby AOG church in Mt Gravatt. A few months later, Denise was then drawn to another five to six thousand strong larger Pentecostal Church called Christian Outreach Centre (COC) nearby in Mansfield. The minister was a man called Clark Taylor, who had a dynamic healing and preaching ministry. Jackie and Nicky initially went to COC school, but they weren't happy there so I put them in local state public schools. They weren't happy there also but happier than at the COC school.

I initially went to the COC church with Denise, but after about one year I decided to return to the Catholic Church mainly because I found that the Pentecostal Church was judgmental toward those whom they considered unsaved. Also one service that I attended, Pastor Taylor made the comment that those who were religious and not spiritual should leave the church in the middle of the service. To me that came across as very disrespectful. Of course my return to the Catholic Church was considered a retrograde step spiritually speaking. But I found that the Catholics were non-judgmental, and I felt a peace being back there. Naturally, my daughters must have been, understandably, confused.

Research at the Mater Hospital

So all in all, everything seemed to be unravelling. But God has His timing, and it isn't ours sometimes. It all started when I had a silly idea that came out of left field. It was that I should measure ascorbate in urine in recurrent stone formers (RSFs). These patients formed about six calcium oxalate stone episodes per year. The rationale for the research idea was that if these folk had more urinary oxalate excretion, then their urinary ascorbate should also be increased because ascorbate has the potential to form oxalate. For this reason, I thought that the idea was a bit too simplistic. Nonetheless, I undertook this work with doctors Jenny Brown and David Cowley from the Mater Hospital.

Well to cut a long story short, we found that RSFs had lower ascorbate levels in urine and elevated oxalate. We eventually found that this reversal was due to ascorbate malabsorption from the gut, resulting in ascorbate being converted to oxalate in the gut. Normally, calcium in the gut would bind oxalate and thereby inhibit its absorption, but we found that citrate was also malabsorbed thereby binding calcium and allowing oxalate free passage into blood and eventually urine. So the combination of lowered ascorbate and citrate gut absorption resulted in calcium oxalate stone formation in the kidneys. Therefore, recurrent renal oxalate stone formation in RSFs was really related to the gut malabsorption of ascorbate and citrate. This so-called simplistic idea resulted in a major medical understanding in this area.

It should be stressed that for normal healthy people, ascorbate is completely safe to take as they do not have malabsorption issues. However, in RSFs we recommended that their daily oral ascorbate be less than 200 mg/day. The minimal daily intake recommended for ascorbate is about 60 mg/day. RSFs were generally advised to drink plenty of fluid because of their condition. Many elected to drink juices rich in citrate and ascorbate, thinking this would be medically beneficial for their condition. But this would have the opposite effect hence their recurrent stone episodes. Those who drank beer or water would have been better off. The renal stone project also had the capable help from Brett McWhinney, a master's program student and Dr Barry Ioannoni then a medical student in training.

While all this stone activity was happening, I met a young immunology specialist from a nearby hospital, the Princess Alexandra Hospital. His name was Ian Frazer and he was a medical specialist from Edinburgh University. At about this time in 1983 a new viral infection called Human Immuno- deficiency virus (HIV) was infecting mainly homosexuals in Los Angeles and heterosexuals on the African continent. This virus, over a period of time, shut down immunity in the infected individuals. This immune reduction was gradual and progressive over many months going from healthy HIV-positive patients to those with lymphadenopathy syndrome (LAS) with night sweats and recurring infections followed later by full-blown Acquired Immune Deficiency Syndrome (AIDS) in which the immunity was almost zero with the patients dying of multiple opportunistic infections.

Because of my PhD work on thiopurines, I became interested in an enzyme called 5-ectonucleotidase (NT) located on the external surfaces of cells including immune active cells called lymphocytes. NT seemed to be important to the development of immunity in newborn babes in that failure of an increase of lymphocyte NT activity after birth resulted in lowered immunity in the newborn. The reason for this is unknown. One mechanism may be nutritional in that NT catalyses the conversion of extracellular mononucleotide purines to their respective nucleosides for absorption into the cell (nucleotides aren't absorbed into cells as well as nucleosides). I am sure that there are other mechanisms which may come to light later.

With the help of my honours graduate assistant Ms Cavelin Hare, we developed an assay for measuring NT in lymphocytes. Dr Frazer supplied the blood from his HIV-infected patients as well as bloods from healthy persons free of HIV infection. The results indicated that HIV infected but healthy patients had 2–3 as much NT activity as those with LAS and ten to twenty times the activity of AIDS patients. So this was a significant finding that showed with depressing immunity, NT reduced dramatically. What was surprising was that the healthy HIV- negative cohort had twice the activity of HIV-positive but well patients. We could not work that out at that time.

We also continued this study in small for gestational age babies as they experience depressed immunity. Several other purine-metabolizing enzymes were also depressed in this group. This work was done mainly with a master's program student, Ms Judith Renouf and Professor Y H Thong – head of child health.

Ian Frazer continued to do research on human papilloma viruses (HPV). Some subclasses of these HPV were shown to be responsible for cervical cancer. Ian developed an HPV antiserum that proved to be effective against the development of cervical cancer. This is now marketed as Guardisil and is now given to prepubescent girls as a protection against the development of cervical cancer later in life. Ian was made Australian of the Year around 2014 for this work, a well-deserved accolade.

During this time I made friends with Professor Y H Thong, head of child health and an expert in asthma control. He took my daughter Jackie under his competent wing and was able to control her asthma. At the same time, we did some research on small for gestational babies as well as looked at biochemically how certain anti-inflammatory drugs were affecting intracellular molecular messaging.

At Christmas 1983, Denise, Jackie and I returned to Adelaide for a short holiday. We stayed with Denise's mother one week and spent the remaining week with my Father. Whilst at Dad's I worked on his garden and repainted some of the house. Dad seemed to have softened over the last three or so months, and we were told that he had had a touch from the Lord. About mid-January we flew back to Brisbane, and the following May, Dad died of a heart attack. I missed him greatly as he was a wonderful dad. Knowing that he was now with the Lord was a great comfort to our family.

In June 1987, I presented my AIDS research with Ian Frazer at the international meeting of Clinical Biochemistry in Den Hague Netherlands. It was well received and I received a prize of 600 Dutch guilders. I was also invited to present this work in Poland but already had made a commitment to present some of the purine enzyme and renal stone work in London. Also my uncle Roland Chalmers and his wife Helen lived in London, and I had wanted to meet and stay with them.

I gave about three research presentations in London hospitals, which were well accepted. I also visited science museums, the National Museum, Tower of London and other wonderful sites. I was in London about five weeks but halfway through met my cousin John Chalmers and his lovely wife Marie-Ange and their three young children Stephanie and the twins Sarah and Julian. John had worked and lived in Germany for many years. I was amazed that the children all less than 10 years old were fluent in English, German, French and Dutch. They were all very kind and we visited lovely villages dating back to the sixteenth century. In one Sunday afternoon drive we visited four countries, France, Netherlands, Germany, and Belgium. By comparison, Australia seemed so large and empty and monolingual. I spent a memorable four days with my German family and grew to love them all.

So as you can see my prayers for research at the Mater had been answered more than one could have imagined. I even ended doing TV interviews on our research, and this pleased the hospital hierarchy as it helped with their fund raising and gave the hospital a good public image.

During this time my daughters Joanne and Nicole had been accepted for nursing courses at the PAH and Mater respectively. Joanne got pregnant out of wedlock and on 25 January 1987, my beautiful grandson Joshua was born at the Mater. Denise and I were more than happy to help Joanne. The father of Joshua just didn't want to know about his son. Unfortunately, the loss was his.

About May 1988, I was offered a chief hospital scientist position at FMC but this time in the department of haematology and genetic pathology. My mother wanted me to return to Adelaide as my father had died of a heart attack in May 1984, and she needed my support. At this stage the research was going so well, and I said to God I was not returning to FMC unless there was a clear directive from Him. Coincidentally, at this time, Denise developed, what looked like, a large shiny white pimple in the middle of her back. I took her to the Mater to have it looked at thinking that it was a benign growth. My boss Dr John Bell called me into his office and told me that Denise had a Clark level 5 amelanotic melanoma. This is a highly metastatic cancerous growth and anything above level 2 is bad news.

Return to Flinders Medical Centre (FMC)

Rightly or wrongly, I took this as a sign from God to return to Adelaide. The melanoma in Denise was found on Friday, and I was going to ring FMC on Monday to turn down the job offer. But this time I accepted the job much to the dismay of my family and some of my work colleagues. In faith, I returned to Adelaide, but it wasn't easy. Denise and I returned with Joanne, Jackie and Joshua. In July 1988 I started work back at FMC, looking after a section called biochemical haematology. Nicole stayed in Brisbane to complete her nursing, and she had become involved with a lad called Andrew.

The work went well but it was a huge wrench leaving the Mater. I published some papers whilst at FMC on mainly Mater- related research. My relationship with Denise deteriorated back in Adelaide, probably not helped by the stress in moving my family. The stress levels escalated and in 1990 I left Denise and went to live with my mother. On the positive side, I was able to help Mum with her heart medication and the specialist confirmed that my giving her drugs at the correct dosages helped improve her health.

The breakdown of my family was very sad, and I wept on many

occasions. I saw Denise about three times per week and helped with bringing up Joshua. I also helped Denise with medical matters when she needed help.

About mid-1990, Mum collapsed and was taken by ambulance to FMC. Her blood pressure was sky high due to poor renal perfusion. This was causing her to have minor strokes or transient ischemic attacks (TIAs). Anyway I agreed to her having her kidney's circulatory system repaired. It was a major operation, but she survived and her BP normalized. Because Mum was now incontinent and couldn't walk due to poor knees, I got her into a nursing home. It was a pleasant home, and she was monitored and well taken care of by nursing staff.

Towards the end of 1990, Denise was diagnosed with metastatic melanoma, which had now invaded her lungs. I returned to the family home to take care of her after a year's absence. She was very involved with her Bethesda church and was very happy there. She was quite mobile, drove her car, went shopping, and visited family and friends. In other words, she lived a very normal healthy life. One night in mid-1991 she complained of diaphragm pain. I prayed and the pain left. I told God if he was going to take my wife please to do it without pain. Well she never had any pain from then on. Praise God.

In 1991, Jackie my youngest went on a cruise with my mum to the Pacific islands, Vanuatu and New Caledonia. On board she met a young man called Craig and subsequently became pregnant. She went to live with him in Melbourne, but it didn't work out and so she returned to Adelaide pregnant. Subsequently, my grandson Matthew was born on 29 November 1991. Denise and I supported Jackie as much as we could. Matthew, like Joshua, was a real blessing to our lives.

On Christmas day 1991 we went to my brother Ricky's house to celebrate; it was a fun day with lots of laughter, food and drinks. I had taken the Christmas week off from work and on 29 December, Denise and I spent the whole day chatting and eating cherries. There were no interruptions, a lovely happy day. In the evening, a Sunday, Denise went to church. She wore a lovely blue dress and a pearl necklace. With her new hair permed, she looked a million dollars as she drove off. Later in the evening, my daughter Jackie phoned and asked if I would babysit my new one-month-old grandson Matthew whilst she went out with her friend. I agreed and took care of Matthew.

I arrived home about 1am in the morning and found there were police waiting around. My immediate thought was that the neighbours were being monitored for having a noisy party. The police however approached me and said my wife had collapsed after church and was in intensive care at FMC. I immediately knew then that her life was over. Apparently Denise drove to a friend's home after church and was having a drink of coffee with her when she collapsed at her dining table. Her friend rang the ambulance, and they took Denise to FMC. I was in shock because she looked so well driving off to church.

The doctors at ICU informed me that Denise had three tumours in her brain, metastases from the melanoma. I sat and spoke to her in ICU for about two hours, but she was unable to respond although there was a slight smile when I shared how I had changed Matthew's dirty nappy. I returned home about 4 a.m. and tried to sleep. At 6 a.m., I returned to the ICU and started notifying my daughters, Denise's mum, her sisters Margaret and Shelly. Denise didn't want anyone to know that she had a cancer, thinking that her close family may undermine her faith that God was going to heal her. She had told my daughters about the cancer two months before, but because she looked so well, they didn't think that cancer was going to kill her. I had to keep this knowledge even from my work colleagues. Understandably her family and mine were in shock at the news of her collapse.

Nicole immediately flew home from Brisbane to join us. At about 6 p.m. on 30 December 1991, I asked the ICU doctors to turn off the ventilator. We then all joined hands and I thanked God for Denise's life and for the blessing she had been to all our lives. As we prayed, her spirit went peacefully to be with the Lord. When I drove home from the hospital, our neighbours were looking forward to us celebrating New Year's eve. When I told them that Denise had died, they were in complete shock. My daughters and I spent a very quiet time New Year's eve 1991.

The next few days were a complete blur as I had to arrange funeral, notices, ministers and everything else involved with a funeral. I had to move quickly because of the holiday season. It was a wonderful service put on by her church Bethesda. Denise had a farewell from planet Earth on 4 January 1992. Even though she wasn't healed from her cancer, I am ever grateful to God for taking her without any pain. She was active until the very last minute. This in itself was a healing event.

Well for months after the funeral, I had a huge fits of guilt for the way that I perceived I had neglected Denise. I didn't go out and didn't want to. Grief and guilt had literally completely run over me like a ten-ton truck. I was in huge emotional pain, and it was only work that kept my mind distracted. Five months after her death, there was a notice that a grief course called Beginning Experience (BE) was available through the Catholic Church I attended. I immediately applied but was told it was too early post death to do this course. I wasn't happy about this thinking the grief counsellors a heartless bunch.

At about ten months post death (October 1992), I was contacted by the BE group to do their course at St Paul Monastery, a retreat property near the Adelaide foothills. I initially said yes but later thought that I was feeling much better and didn't need to do the course. Despite desperate efforts to cancel this course, I wasn't able to contact the BE group and was basically forced to waste a weekend in October, or so I thought.

At the course there were about 100 grievers and 12 counsellors of different denominations involved. So we were broken up into groups of about 8. Everyone in these groups shared their grief. Losses were due mainly to death, divorce or separation. The course ran on Saturday and Sunday, and we slept in at the monastery Saturday night. On Saturday night, I could feel myself getting more and more emotional, almost in tears. I tried to escape Saturday night when everyone was asleep but couldn't, in the dark, find the way out of the monastery to my parked car.

Sunday arrived and for our final task we had to write a letter to our loved ones and read it to our group. Well I was a complete disaster and bawled my eyes out reading this letter. Just note that I am not one to burst out crying, especially in public in front of people I don't really know that well. But Sunday afternoon the floodgates burst. As a result of this event, I was able to get on with my life though at that time I thought myself pathetic. Well everyone hugged each other, and it was very much a healing weekend. The day ended with a mass and holy communion. When I got home Sunday evening, I expected the house to be very smelly as I had left the dirty washing on the floor of my closed bedroom. What hit me was completely unexpected because the room smelt of roses. I suppose it was God's way of approving of my grief weekend and my farewell letter to Denise. I didn't go to work the next day, as my eyes were red from so much crying.

After this weekend I started dating some of the other women that I had met on the course. It was purely platonic, and we shared some good times going out one to one or as a group. In February 1993, I had a wonderful bus tour of New Zealand, going by myself. In August 1993, I cooked some Indian food for Joanne's friend, a medical student at FMC. Whilst at her party, I met a lady whom I shall call Wendy. We got on very well and I started dating her. She was seventeen years my junior, but when asked, she said that our age difference didn't worry her.

We fell in love and in 1994, I sold my home and moved to an outer area of Adelaide called McLaren Flat, about twenty kilometres from FMC. Wendy sold her house and moved in with me with her son and two teenage daughters. In my naivety, I was hoping to bring our two families together, something like the Brady bunch. Putting it mildly it didn't work and Wendy moved out to rented accommodation after a few months. I loved her and still visited her, and in June 1996 we married and bought a house together in Seaford Rise. Her youngest daughter lived with us, completing her high school education.

MY PHYSICAL AND SPIRITUAL JOURNEY INTO TRUTH

In 1997, I was invited to speak at an International Conference in Udaipur in the Rajasthan region of India. Wendy wanted to visit India, and so we both went and enjoyed our trip there. After the conference I spoke at Jaipur University and then flew onto Delhi. We toured around Delhi and then caught a bus to Agra where my mum grew up and was educated. We visited the Taj Mahal and Akbar's tomb, both wonderful experiences. We then took a taxi 400 km from Delhi to Naini Tal where I had grown up till age 9. Naini was absolutely beautiful, more like a Swiss village. I found my old school, St Joseph's, and the headmaster kindly showed us around this impressive old school built on British lines. He also showed records of when my father, uncle and I started and finished at this school.

How we survived the return taxi drive from Delhi to Naini and back, I will never know. Our angels must have been working overtime on this trip. Let us just say that Indian drivers have never heard of road rules. All in all, we had a wonderful few days in Naini.

Wendy and I separated in April 2000, and I rented a one-bedroom flat while she stayed in our jointly owned house. The reason for the separation was basically due to friction between her and my family. I wanted us to have counselling to repair our marriage and had arranged for us to do so. But Wendy said I wouldn't listen to any counsellors and therefore viewed the whole counselling thing as a waste of time. I guess that I was partly to blame, as I probably hadn't resolved all my grief in losing Denise. In September 2000 I bought a home in Aberfoyle Park, only ten-minute drive from FMC. This was handy as I used to be called out in the middle of the night to troubleshoot specific assay problems. I still live in this house 19 years later.

Our Stress Work

Whilst all these marital problems were going on, I was approached by Jane Blake-Mortimer, a psychology student from the University of Adelaide, wanting to do a PhD on any possible biochemical changes perhaps responsible for immune depression in patients with chronic psychological stress and depression. For reasons unknown, other research colleagues had referred her to me. My initial response was that I couldn't think of anything immediately. But then shortly after, I recalled the earlier NT research on the homosexual population at the Mater hospital with Ian Frazer and suggested that we could measure lymphocyte NT activity in these depressed patients and see if they differed to a normal healthy cohort.

Initial studies done as a short blind trial with few patients indicated that NT activity was about 20–30 per cent lower in depressed/stressed clients when compared to a healthy control group. Before starting this work on a larger group of patients, I had to get approval from the head of my department. It is strange how God works. My boss was friendly with Jane's boss, and my boss's son was going steady with Jane's boss's daughter. And so the NT work was approved, praise God. Otherwise it would have never got off the ground.

The initial trial of depressed NT in stressed patients was confirmed in a larger cohort of patients. Further investigations by us showed that the depression of NT was caused by an increased free radicals generated in stressed persons. Giving patients antioxidant mixtures of ascorbate, vitamin E and coenzyme Q10 normalized NT in these stressed patients. Also we found antioxidants prevented the damage of NT by free radicals that we had generated in test tubes. The results suggested that antioxidants given to stressed and depressed persons could help enhance their depressed immunity. However, the antioxidant vitamin mix beneficial to immunity needed to be further worked on. Ascorbate on its own was ineffectual in protecting NT even at a high dose of 0.5 grams/day for three weeks.

The result of this study also helped explain the 30 per cent depressed NT in HIV positive but well patients that Ian Frazer and I had worked on several years previously. Obviously the HIV-positive patients would have been depressed and stressed, but psychologists hadn't confirmed this aspect at that stage.

From the depressed patient work done with Jane Blake- Mortimer and the findings with Ian Frazer, one would expect that antioxidant administration would be beneficial to long-term immune health. The question, however, was which antioxidant combination would work best? One had to bear in mind that too much may negate free-radical-mediated microbial death by cells such as by neutrophils and macrophages. This immune response is part of what is called the innate response and is the first line of defence against microbial attack. The other innate response cohorts would be skin and saliva.

The response that NT was indirectly measuring was the acquired immune response. This is the big-daddy response that is specific against certain antigens on bacteria and viruses. And it is also active in fighting the development of certain cancers. The acquired immunity also has a memory aspect in that it will remove, even more efficiently, any of the same nasty microbe that may try to attack us a second time, even years later. This is the principle on how vaccination works. That is a non-virulent microbe (or dead one) is injected into us, and the acquired immunity springs into attack mode. Should living bacteria, related to the one used in immunization, try to enter our bodies (even years later), then it will be attacked immediately by the primed immune response.

Well you may ask how does stress cause a high pro-oxidant state resulting in damage to the acquired immunity. Here I have to theorize. Basically I think stress somehow dampens the acquired immune response perhaps through the increased release of cortisol, a stress hormone. The innate response, to compensate for the depressed acquired response, fires up to compensate resulting in the increased generation of free radicals which then further damages the acquired response through shutting down NT and possibly other important differentiation ecto-cellular markers. So the normal homeostatic balance between the innate and acquired responses is thus put out of balance by chronic stress and resultant depression arising from it. You may ask me if I take antioxidants. Before and while this work was in progress, I would have said no mainly because I didn't know which ones to take. Also I am not a big pill taker.

But it is funny how God works behind the scenes. Let me explain.

When this work with psychologists was about 90 per cent done, about July 2001, a strange thing happened. Remember that I had left Wendy my wife in April 2000 and my mother died in July 2000. I had spent three days in the Flinders Ranges bushwalking with my 13-year-old grandson Joshua. On the final day after a strenuous hike, it was time to return to Adelaide about 450 km south. I drove whilst Josh slept on the back seat of the car. About 50 km into this trip, the road ahead suddenly seemed distorted. I closed the right eye and all seemed well. However, when I closed the left eye, my central vision went black. Well I drove on and dropped off Josh at his mum's home and immediately made an appointment with an eye specialist the next day.

Briefly the specialist diagnosed that I had macular degeneration (MD) in my right eye. Basically the veins at the back of the eyeball had proliferated and damaged each other resulting in blood loss at the back of my eyeball or the macula. My mum had it in both eyes and was probably due to her being a smoker for over thirty years. The specialist said that I should take antioxidants to protect the remaining eye and referred me to a New England Journal of Medicine article on the protective effects of antioxidants on the macula and eye health in general. Not wanting to lose my remaining vision, I started taking antioxidant pills called MacuVision. Since taking these pills for over 18 years now, I very rarely get sick. I also take multi B vitamins as they are reputed to have anti-inflammatory effects. In addition, I take Flomaxtra (0.2mg/d) for my prostate (I have benign prostatic hypertrophy or BPH) and low dose Nexium (10 mg/d) for mild gastric reflux. I also take an herbal mix, called Prostalobium (1 tab/d) reputedly good for prostate health. I am fully aware that one cannot form firm conclusions on just one person's hearsay. I wasn't expecting a low-sickness rate in myself, and I don't have shares in any of these drug firms but think that there is good enough reason to do a long-term follow up on rate of infections in a large cohort of persons taking MacuVision or antioxidant products similarly related to it.

I think that my MD was caused by stress due to grief from Denise's death, my separation from Wendy and my mother's death. My NT levels were low as well, confirming biochemically my high-stress level. The free radicals must somehow have had a role in causing red-cell proliferation behind my eyeball. My remaining eye is still well 18 years after the initial event. Praise God. An eye specialist recently confirmed that I have the dry form of MD (as opposed to the wet form) and that nothing can be done in this circumstance.

This stress work mainly involved doctors Jane Blake-Mortimer, John Hapurachi and Tony Winefield.

Xanthochromia Work

Whilst at work one day, I prayed to God that I would just love to know if any one of my test results really did save a human life. In the lab we would do all these tests but never get any feedback from the clinicians as to whether our tests were helping anybody, hence my prayer.

My lab does a plethora of tests like HbA1C measurements for diabetic control, folate, ferritin and vitamin B12 as factors in anaemia, detection of haemoglobinopathies and genetic testing for mutations in persons predisposed to getting bowel cancers.

Within a month of saying this prayer, a sample of cerebrospinal fluid (CSF) appeared on my desk after my lunch break. The request was for xanthochromia detection. As I was sometimes asked to do unusual assays I had assumed this was a specimen that I had to analyse. Looking up my medical dictionary I found out that xanthochromia (XC) was a yellow colouration in CSF due to bilirubin resulting from a bleed leak, presumably from the brain. It should be mentioned that bilirubin is derived from metabolic breakdown of haemoglobin, the red colouration of red blood cells. Hence, a brain bleed would result in elevated CSF haemoglobin and eventually bilirubin.

So I did a quick spectrophotometric (spectral) scan of the CSF and found a peak indicating bilirubin. I reported this result to the neurologist (Michelle Kiley) who informed me that the specimen had erroneously come to my lab instead of the microbiology lab. On further questioning, I found out that the microbiologists did this test by holding the CSF up to the light and reporting the absence/presence of yellow colouration. I notified the neurologist that the eye test was not as sensitive as the spectral test. The neurologist immediately became interested in the spectral test and informed me that the XC was used to detect sub-arachnoid haemorrhaging (SAH) or brain bleeds. Sometimes they also used CAT scans to try and detect SAH. Apparently SAH indicates a bleeding aneurism which if untreated would result in death. So this test is an urgent one that needed to be done immediately.

The neurologist then started sending me CSFs from non- bleed and suspected SAH patients. The spectral method turned out to be very sensitive in picking up SAHs, and we subsequently published this work. And the method became routine in our lab and many others around the world. I was initially unaware of the significance of this method until I gave an oral presentation of the method at the annual meeting of the Australian Association of Clinical Biochemists in Brisbane. There was considerable interest in the spectral method, and I was awarded first prize ($1,000 plus plaque) for the most significant presentation at this meeting by Roche Pharmaceuticals.

Well a month or so after this presentation, Michelle Kiley the neurosurgeon came into my lab and notified me that my test saved a 34-year-old man's life. The CAT scan was negative, but my test indicated SAH. A follow-up test, where dye is injected to highlight the brain vasculature, indicated that indeed there was a bleeding aneurism. The doctors were going to send the man home that weekend after the CAT scan, but because of my test, the aneurism was operated on to stop the bleeding. So my prayer to God had been answered. Praise God. And that was not the end of the story.

My European Trip with Krista Marcher

A female scientist named Krista Marcher, from Denmark, contacted me re the XC test. Being about my age and through the email contacts, we became friendly. She had worked about twelve years in Paris and offered to show me parts of Europe during her holidays in August to September 2001. Because of my grief due to separation from Wendy in April 2000, I was reluctant to go. However, through the urging of my brother Mark, I decided to take this European holiday with her. Before embarking on this trip, I had one stipulation that our friendship would be platonic and that we would sleep apart.

Well, I met Krista at Copenhagen airport. She turned out to be a lovely and friendly person as I had expected from our emails. We caught a train to her hometown of Esbjerg and after a few days embarked on a train trip initially to Germany. In Germany we met my cousin John Chalmers and his lovely wife Marie Ange. It was great to see them once more, including their children Stephanie, Sarah and Julian, now all grown up and adults in their twenties (remember that I had met them fourteen years earlier when they were small children). Krista fitted in very well with the family, especially

Marie-Ange as she could speak to her in French. Marie Ange was born and grew up in Belgium. Remember Krista had worked in Paris for a dozen years before settling back home in Denmark, hence her fluency in French. In Germany, we visited many lovely old towns dating back to the fifteenth century.

After a great week with the German Chalmers clan, we trained on into Paris. In Paris, we saw many museums, art galleries (Louvre) and met some of Krista's friends. Not being able to speak French fluently made one feel a bit isolated, but Krista always filled in all the many gaps. At the urging of her friends, we saw the Palace of Versailles and after caught the train to Aix en Provence. The journey from Paris to Aix was 800 km but took only three and a half hours in the TGV train that travelled at 320 km/h. We stayed about five days in Aix, a delightful little town with narrow cobbled stoned streets and visited some lovely art galleries including the home of Cezanne, the famous artist.

While at Aix we went by train to Cote d'azur and visited lovely little towns like Cannes, Monaco, Avignon, Arles and many other smaller ones. On the way back to Paris, we dropped into Marseille. We returned to Paris and caught the train to London.

Since I had been to London before, I showed Krista some of the sights like Buckingham Palace, Westminster Cathedral, and 10 Downing Street. Whilst visiting St Paul's Cathedral, there were prayers being made for the people killed by terrorists. The prayer turned out to be a sung mass with the archbishop, priests and a huge choir. I said to Krista this must be some terrorist attack, not having seen the news or read any newspapers. We were unable to buy a newspaper as they were all sold out. That night we watched the TV news and saw the 9/11 attack of the Twin Towers in New York. We were in London on 11 September 2001 and were absolutely shocked by what had happened that day. A massive tragedy which stunned us and I am sure, many others.

After London, Krista flew back to Denmark and I caught a bus to Leeds where I had planned to visit my son-in-law's (Matthew's) mother Audrey. I spent a memorable five days at Audrey's, being shown York and other lovely towns and more castles than I can remember. I also met Matthew's sister, her husband and her children. Audrey really spoilt me rotten. After the visit, I caught a train to Stansted airport near London and flew onto Esbjerg to rejoin Krista.

In Esbjerg I visited Krista's hospital lab and set up the XC test. We also did some lovely visits to some old Viking towns. After four days we went to Copenhagen by train, and Krista showed me around this beautiful city. The most memorable was the Tivoli showground where Walt Disney got his ideas for setting up Disneyland in Los Angeles. After about two days in Copenhagen, I flew back to Adelaide. Krista had been a lovely companion and very kind in showing me around. I don't think I was much fun to be around, still carrying grief both from my split with Wendy and from the deaths of Denise and my mother.

Return to Adelaide in 2001 after the European Holiday

When I had returned to Adelaide, Wendy rang and I told her of my European holiday with Krista. She was furious, thinking that I had had sex with Krista, which I hadn't. Despite my assurances, she refused to believe me, and a week later, I received divorce papers from her. This caused me much pain; however, what could I do? And that was the end of that. For the next two years I continued my research into the biochemical changes in stressed/depressed persons, and we published in this area. On 10 August 2003, about five days after my 65th, I took a retirement package. However, I continued my stress research and teaching for the next five years.

In about 2003, I met a nice Indian lady called Mavis. Mavis was a lovely lady, and we became good friends with a common history in India. She came from a large Indian family with twelve siblings, and we used to enjoy partying together. Mavis and I did dancing lessons, and we danced quite well together, especially doing rock and roll, our favourite dance. We also holidayed together, showing Mavis parts of Melbourne, Sydney, the Gold Coast and Mt Gambier. In 2008 I met Leonie and we fell in love. We have been together till now 2019.

My meeting Leonie was quite God-ordained. Despite my friendship with Mavis, in 2008 I felt ready for a deeper relationship and prayed to God about it. I also told Mavis about my feelings though they did hurt her. I believed in being honest with her beforehand.

There is a newspaper called the *Messenger* that covered local news such as council decisions, locals doing well in sport etc. Normally I would pick up this newspaper on Wednesday morning and throw it directly into the recycle bin since the news didn't interest me. However, one day, the paper was wet from overnight rain and so I hung it over the kitchen chair to dry before throwing it out. A few days later whilst having breakfast, I thought that I would look at the now dried *Messenger* paper. There was a section where people indicated that they wanted to meet others of the opposite or the same sex. I must emphasize that I had no intention of meeting anyone. However, I saw an advert by a woman wanting a male golf partner. Being a golfer I thought that might be nice. The lady golfer was Leonie and I subsequently arranged to meet her a week later in a populated area on Jetty road Glenelg a suburb of Adelaide. I was so nervous. We had lunch at a local hotel, and I found that we were immediately at ease with each other and got on really well. At that first meeting we talked away about our respective life's journey. I liked her straight away and was hoping that I might meet her again. A few days later she rang saying she needed a minor operation and then was then going to meet her friend Sue in Melbourne. My initial reaction was that she had met a suitable other better than me and maybe was letting me down softly. I was feeling disappointed but 'C'est la vie,' that is life. About one month later, she rang me, and I rang her back. She had been honest about her operation and the Melbourne trip. We arranged to go to a play on our first date. We felt at ease with one another, and on that date I kissed her. I drove home that night with my head spinning, thinking what was I doing. Well the next week we went to a movie and the rest was history. We fell in love and yes, we played golf. On the golf course she was very competitive much to my shock because on the movie and dinner dates she was so sweet. She not only beat me at golf but also told me where I was going wrong in my swing. I learnt a lot about humility on the golf course with Leonie.

Spiritual Aspects of My Journey in Life

Throughout my life, I was literally unaware that God was directing my paths. However, looking back it has become quite obvious that God was active in my life. I shall try and describe my revelations in this area of my life.

I remember as a one- or two-year-old being stung by a scorpion that had fallen into my pram. The pain was intense and the memory, though faint, still remains with me even till now. As a six-year-old in Naini Tal in India, I fell into the lake one day but was rescued by my aunty Clare, who was only about 12 years old at that time. She carefully dried my clothes, and we kept the incident from my parents who would have been very angry at Clare's lack of proper supervision, if they had found out.

At the age of eight or nine I used to ride my bike to an isolated area near Delhi airport to see the DC3 airplanes landings. I was unaware that there had been a lot of angst by Indians against the British rule in India for so many years. One day at this plane viewing, I was chased by two Indian men who were brandishing knives. I rode my bike at full speed and just managed to get away. Needless to say, my plane viewing came to an abrupt end after this incident.

MY PHYSICAL AND SPIRITUAL JOURNEY INTO TRUTH

Later in life, I realized that God's hand was there, protecting me in these incidents.

I rode motorcycles for close to forty years and came off many times with only minor gashes, back in the days when helmets weren't compulsory. God, I felt was protecting me all along because I drove my bike very recklessly especially in my teens and early twenties. God also protected the Barr Smith Library in Adelaide University from burning down when my bike caught fire one night.

When we came out from India, I was about ten and my brother Geoff was about 7 years old. We used to go down to the beach, which was located about 100 metres from our home and play in the sea from dawn till dusk. We couldn't swim and could have easily drowned but for the protection from God. In those days, people from India including my parents didn't see the importance of learning to swim. Eventually we taught ourselves how to swim because there were no swimming lessons available then.

All the research and projects that I had undertaken in hindsight I could see God's hand in directing my life. For example, I worked with my excellent supervisors, which helped me get a PhD. My cancer work, IMVS projects, renal stone work, AIDS work, the stress work and the XC method all had the hand of God on them. During my research, I met about four Nobel prize winners, all associates of Professor Chev Kidson. One had won the prize twice (Dr Fred Sanger) for sequencing protein and then DNA. He related to me how much he loved building and sailing yachts. In short, I went from a near-school dropout to working on top medical projects, all directed by God. Also my ability for scientific/medical work all came from God, so I really couldn't really take any credit for whatever achievements I may have made.

I just want to emphasize that we are all important in the scheme of life. God gifts us all in many different ways. My brother Geoff, for example, was gifted technically and could build home additions, fix cars etc. I felt quite inadequate when I compared myself to him in this area. In this book, I have described how God has used me in my life as this story is about me and not because I am better than anyone else. God forbid!

Spiritual Lessons Learnt

Faith is simply blind belief.

It is generally believed that faith seems to be belief of something that one cannot see or touch. That is, to the religious, it is a belief in an invisible spiritual world. We cannot see nitrogen gas, but it can be cooled to become a liquid and then we can see it. One, however, cannot visualize the spiritual world by human endeavour.

Christian faith is not blind belief and so is vastly different to other faiths. Paul's letter to the Romans in the first chapter tells us that the creation reflects the work of God so that nobody has a reason to disbelieve. That is what is seen reflects the invisible or spiritual workmanship of God. As a scientist working at the molecular and cellular level, I can certainly concur that there is loving, super-intelligent being out there who has created our world with its billions of teeming life types or kinds. And the Bible clearly tells us that this creator is Jesus Christ (John 1:3, Coll 1:16). Why must we invoke Christ as the only manifestation of God? First of all He made the incredible claim that He and God the Father were one (John 10:30). His words and miracles testified to this fact and also the fact that Christ rose from the dead and is now alive and working in His believers through the Holy Spirit. This claim of being God

He repeated throughout John's Gospel, and it eventually got Him crucified for the sin of blasphemy.

In addition, numerous prophesies about the coming Messiah were made hundreds of years before they were all truthfully fulfilled in the person of Jesus Christ. Christ's rescue of humanity from the curse of sin was prophesized by God Himself way back, almost in the beginning of time (Gen 3:15). So the Christian faith is grounded in creation, history, prophecies about Christ and the actual appearance and documented miracles of Christ on our planet about 2,000 years ago. So the Christian faith is far from being blind. No other faith on earth can say that about their god or gods. The discussions following below will confirm the Godhead of Jesus Christ.

Comparisons are odious.

We should never compare ourselves with anyone else. God makes us unique. As the saying goes, He threw away the mould when He created us. I will never be a Tiger Woods at golf, but maybe I am better at science than him. Unfortunately medical research is not a spectator sport!

Comparisons cannot only be with other persons but against other belief systems. For example, I have met many anti- Catholic Christians who can only see theological faults with the Catholic beliefs but overlook the charitable works done by that church. I personally cannot go along with certain Christian beliefs because they clash with my understanding of what God is saying in His word. However, I am not going to denigrate what they believe. My understanding is limited by my sin nature but certainly not by God who is sinless.

Forgiveness is a must.

The bible emphasizes the importance of forgiving those who have hurt us. I have heard the comment that unforgiveness is like drinking poison. It destroys the unforgiving person but not the person who the unforgiveness is directed against. My experience in this area has shown this to be true. One senior person I knew was very cruel to me at work, so much so that I used to cry out to God in my lunch break. God asked me to pray for that person's salvation. Initially I resisted but eventually relented. My first prayers for this person was through gritted teeth, but with time I softened and the prayers came easier. With prayer this person's attitude also softened towards me. This is an area where we need the guidance of the Holy Spirit. God tells us that we are to love our enemies and even to pray for them (Matt 5: 44). So my prayer didn't change this person, but God, working through it, changed my heart and my love towards that person.

The other necessity for forgiving is that unforgiveness and unresolved anger can open us up to the demonic realm (I deal with this area later on). That is Satan can influence us by throwing us more into a spiral of greater anger. In the extreme this can lead to murder. As I have stated, the key is to pray for our so-called enemies as Jesus has instructed us to do and not to feed that unforgiveness and thereby become more and more embittered (Rom 12:14–21; Matt 6:9–14).

God is sovereign.

That means He has complete control of our lives and everything else. We are here to serve Him as Lord and centre of our lives and not He to serve us. In other words, belief in Christ is not just an add on to our lives. In the past I believed that if I asked God in Jesus name, I would get what I asked for. In other words, God was my servant, an ATM machine and Jesus my pin number. What I came to realize was to ask in God's name was to ask in accordance with His will. As we say in the Lord's Prayer, 'Thy will be done.' God loves us and sometimes He says no because He knows what is needed and best for our lives. It is a bit akin to when we say no to our kids when they want bags of teeth-rotting lollies. However, having said this God does grant us our requests when they are made in a spirit of love.

God has blest us in ways unimaginable.

For example, I see the expression of God's love for us, in and through His creation. When we love someone, we may buy him or her flowers. God has given us incredible beauty in nature for us to admire in the multitude of different flowers, trees, animal life, marine life etc on our planet Earth. Not only that but there are billions of galaxies and stars in the heavens for us to admire. As a scientist I have been privileged to study God's incredible creation at the molecular and cellular level. How in the human cell, tens of thousands of molecules harmoniously interact to maintain normal functions within the cell. How, in the human body, the different organs interact and communicate with each other to maintain our health, healing and normal function. God is truly an awesome God; there are no words grand enough to describe our God. He truly is worthy of glory, honour and praise. Creation really does scream out that God exists and truly loves.

The greatest expression of God's love.

Simply put, when He gave His life for us at Calvary 2,000 years ago, He set aside His divine nature and became man on our behalf with the final intent of reconciling us to God by giving us His righteousness. This rescue by God had to be done because our righteousness based on our good works were like dirty rags and not up to God's requirement and would have kept us eternally separated from God (Isa. 64:6; Phil 3:9). Jesus at Calvary took our sins on His horribly tortured body and replaced it with His holy righteousness so that we could be acceptable to God. His death was a horrendous event and was done for our salvation, out of pure love.

This gift of God to us is called grace, which is an unmerited gift of God to us. As the Bible says we are saved by grace and not works that any man may boast (Eph 2:8, 9). However, the next verse (10) does stress that we are designed by God to do good works. Just to stress further, God has designed us to do good works but salvation comes solely through accepting and believing the loving sacrifice of Christ on the cross for our sins and not by good works alone.

Some on Earth elect not to love since love does put one in a vulnerable situation. As the song says, 'You always hurt the one you love with a hasty word you can't recall.' Usually when I have been treated unfairly by a loved one, I tend to fly off the handle and isolate myself. This is in contrast to Jesus, when on the cross, cried out to God in His love even for His tormentors, 'Father forgive them for they know not what they do' (Luke 23:34). That is a level of godly love way above my earthly comprehension or ability.

God does direct our paths through life.

Because of His great love for us, He is directing us though it sometimes seems that He is far away. However, when one looks back one can see where He has led us. My experiences and career described above is a classical example of God's direction. Obviously in life we have to make moves, but when we do, God can work on us. It is difficult for God to motivate a stationary subject. If the move is not a wise one, God will open another door. It is important always to prayer to God over what may seem like a long time, but then God, in His perfect timing, does things. My experience at the Mater was a classical example.

Initially in the Mater job I was involved in administration of the lab and research was not on the horizon. For about three months I prayed that God would open up an avenue to do research. Then I had a silly idea to measure ascorbate in RSFs. I initially argued with myself against this idea because one expected high-oxalate excretors to have high ascorbate since ascorbate can generate oxalate. However, in my desperation I experimented this idea. Of course stabilizing ascorbate in urine is a problem that God helped me solve. To cut a long story short, we found out why RSFs were having around six oxalate stone episodes per year. We also published about ten peer-reviewed publications in this area as well as presentations overseas.

This initial research opened up further research collaborations in the AIDS and arthritis areas. When I returned to FMC in 1988, I had the opportunity to work with academic psychologists on biochemical changes responsible for depressed immunity in stressed/ depressed individuals. We made some significant findings in this area thanks to God's direction not only for the research ideas but also for bringing the right research psychologists to my lab. God also played a role in getting me to detect early SAH by developing a sensitive XC method.

The word of God, the Bible, is absolute truth.

Holy scripture or the Bible is described as being inspired by the Holy Spirit (2 Tim 3:16) and so one would expect it to be absolute truth, as God never lies. In my early Christian life, I had doubts about it being absolute truth and have covered some of my doubts in the more scientific areas outlined later in this book. However, after much prodding and poking over many years I have come to realize that the Bible speaks absolute truth. Christians that I have met over the years have disagreed with me on this topic. They claim it depends on how one interprets scripture. To me scripture is quite clear in what it says when taken literally. The exception is in the genres of poetry or prose or with metaphors. The result of reading what one interprets from a subjective bias is confusion. For example, Christian ministers have commented that there are many ways to God through the gods of many religions in contrast to John 14:6 where Christ states that He exclusively is Way, Truth and Life and the sole way to God. In John 10:9 He emphasizes this further by saying that He is the exclusive door to salvation. Other Christians have made comments like (1) the God of the OT is cruel whereas Jesus is loving, (2) We don't believe in a heaven or hell and (3) The

Bible shouldn't be taken literally. I don't mean to make judgements, but these comments fly in the face of biblical truths. I guess that we are all on a journey on which God is changing us. And we all need to have lots of patience on this journey, especially me. I feel that as I move along with Christ, I become aware more deeply as to what is truth

In this world we are bombarded by messages that purport to be truth. For example, adverts tell us that we will be fulfilled if we have certain material things like face creams that keep us young, clothes that make us look desirable, cars that will make us happy, foods that will completely satisfy us, diets that will give us the perfect figures, health etc.

There are other areas such as evolution that I have come to realize are lies. I cover this subject later on. I feel that the concept of global warming due to CO_2 gas emissions by man difficult to reconcile scientifically as truth. This is because CO_2 represents only 0.03 per cent of all atmospheric gasses, and of this man is responsible for less than 0.01 per cent. Plants consume CO_2 for energy production and in this process produce oxygen, a gas essential for the viability of humans and biological life in general. Nature has a way of balancing out any excess CO_2. For example, with excessive temperature rises and there is more water evaporation resulting in more rain, which removes the excess CO_2. Also the greatest greenhouse gas is water vapour and is responsible for about 97 per cent of global warming. In short it seems that excess atmospheric CO_2 is a minor player in global warming and that there are many numerous interacting factors, other than CO_2 alone, contributing to this perceived event.

Reducing atmospheric CO_2 in order to reduce global warming by even one degree centigrade would seem to be an almost impossible task for humans to achieve. However, protecting our forests worldwide should have a high priority because they not only help reduce greenhouse gases like CO_2 but also protect animal life against extinction worldwide.

There are other activities that humans can control that can have a serious positive impact on our environment. For example, we can

MY PHYSICAL AND SPIRITUAL JOURNEY INTO TRUTH

and do produce cleaner energy naturally through solar, wave, wind and hopefully fusion if developed in the future. We can also clean up our environment more through greater efforts in recycling and by focusing more on using degradable plastics. Current overproduction of plastic products, due to demand, is resulting in pollution of not only our lands but also our oceans and waterways and the vast amount of life contained within them. Therefore, this overuse of plastics needs to be greatly reduced, and we can achieve that through reducing demand and by other means.

Solomon a king around 1,000 BC was a very wise man. He wrote mainly in Ecclesiastes, proverbs and psalms amongst others. I mention him because of some of the truths he discovered in his journey through life. He also was probably the richest man ever. He had incredible wealth and had over 300–500 wives and concubines. At Solomon's request, God had bestowed him with great wisdom. Kings and queens visited him from afar to experience this remarkable man for themselves. Well what did he think of his great wealth and good fortune? Ecclesiastes 1:2 summarized his thoughts on this matter: 'Meaningless, meaningless! Everything is meaningless.'

In Ecclesiastes 3:11, he said about God, 'He has also set eternity in the hearts of men...' In other words this God hole in us (eternity) can only be filled by God Himself. In the same vein, Genesis 1:26 states that we are made in the image of God and also made to have fellowship with Him (Gen 3:8). We can truly only get a glimpse of God and understand Him more fully through the Holy Spirit inspired revelations of Him in the Bible (2 Tim 3:16). The professor of English C S Lewis summed it up beautifully when he said that God cannot give us peace and happiness without Him central to our lives. Jesus states that you shall know the truth and the truth will set you free (John 8:32). Jesus in John 14:6 also said, 'I am the way, truth and life. No man can come to the Father except through me.'

In science we try to find the truth in what the cause of a certain disease is in order to treat it. In a similar vein engineers and builders need to have knowledge or the truth about the strength of certain materials in order to build a structure such as a bridge, building etc. So truth obviously is a must or essential for us and all areas of our society to work. Unfortunately, in this world many believe lies and one of the most difficult things to defend is the truth and not freedom as many may think.

Church membership, open to all?

My daughter Nicole and I considered this question as at their church. Matt (her husband) had to lead a discussion on this topic. The question asked was can their church allow same- sex couples to worship at their church? Part of the church said it was OK since Jesus often mixed with and came to save the lost sinners like tax collectors and the like. That is the church, like Christ, should be open to all regardless of their lifestyles. Others disagreed because they felt that it was condoning a sinful lifestyle.

Firstly, the church should be open to all. To me church is a group of sinners who recognize their sinful state and have come to God to seek forgiveness through repentance of sin and thereby receive the benefits of the cross of Calvary. That is Christ is the only one who can redeem them from their sins and thereby reconcile them to God (John 14:6). The apostle Paul often refers to their past sinful life of Church members and how Christ has changed them from within and we, as born-again Christians, can all relate to that (Eph 2:13,14,19). Galatians 5:21, 22 demonstrates how we were before coming to God and how the Holy Spirit has changed us from within. Verse 24 describes how we have basically nailed our sinful life to the cross of Christ. In short when we come to Christ, He changes us dramatically from within. Nonetheless we are still sinners to the day we die, and if we say that we are not, then we are calling God a liar (1 John1:10). Christ changes us from within through the Holy Spirit and imparts His righteousness to us to make us right with God. Our salvation and the process of sanctification are totally dependent on Jesus Christ Our Lord and Saviour.

So what should the Christian church allow within its bounds?

What is acceptable behaviour in church membership? Paul in his letter in 1 Corinthians 5 spells it out clearly.

A man in the church had been sleeping presumably with his stepmother, after his dad died. Paul asked the church to talk to this guy about his behaviour and if he didn't give up his sin then kick him out of the church and let Satan deal with him. The guy's sin was basically contaminating the church. Sounds tough but the good news is that he does repent and comes back to God and the church (2 Cor: 7). Paul encourages the church to love and forgive him. So the church does need quality control (just like my lab) otherwise anything goes. The church cannot accept sinful behaviour that God speaks against since it represents the body of Christ. In other words, the church must be on guard not to become secularized. It must be different from the world. The judgement I have spoken about relates to those fellowshipping within the church and not those outside the church; that is the world. Let me attempt to fill this out.

As Christians we attempt to live out our lives according to God's moral code as described in Holy Scripture, the Bible. Those outside in the secular, non-believing world also have a moral code some of which is in accord with Christian beliefs like 'Do unto others as you would have them done unto you' (Matt 7:12) and show kindness and love to those poor and in need (Gal 2:10). However, many of the secular or non-believers uphold morals that contradict the word of God like:

1. Homosexuality and same-sex marriage (SSM) are OK provided the partners love one another. Christian ministers, cake makers and photographers can now be sued if they refuse to offer their services to SSM or label homosexuality sinful behaviour.
2. Abortion is allowable when it interferes with the person's life plans. Christian doctors are being coerced into referring patients for abortion against their Christian beliefs. God's command against killing another human is very clear.

3. Gender is fluid and you can nominate whatever gender you feel you are. All that is needed sometimes is for a male to just wear a female dress. This has put women at risk when trans females end up in female shelters, change rooms, hospitals, prison wards, and toilets. Also trans females competing against females in sports have a distinct advantage over females because of their larger muscle mass. Gender fluidity is being taught to young children in many schools around the world despite the fact that it was shown by psychiatrists to be mental condition that can respond to appropriate counselling without the need for hormonal and surgical interventions. God's opinion on this matter is quite clear in that He stated making them male and female from the beginning of time (Gen 1:27). The secular world would not concur with this belief.
4. Laws are coming in to decriminalize prostitution. It is felt by many that this will make women more open to abuse in this profession by pimps coercing them. The Nordic model seems a better law in that it prosecutes users of this work (pimps, customers) and protects women. It also gives women an avenue to leave the profession if they so wish. The Nordic model has now been accepted by several countries around the world.
5. Conversion therapy involves helping those who want to transition from homosexuality to heterosexuality and those with gender confusion. The therapy has been shown to work well and yet is banned in many countries.

Unfortunately many of these secular ideas are taking root in many Christian churches around the world. These churches are, as a result, losing their saltiness and being a light to the world (Matt 5:13-16). As already mentioned problems arise when one reads, usually from a subjective bias, something other than truth contained in the word of God. I have met Christians, who don't believe in heaven or hell, the incarnation of Christ, Christ as God, Christ the only path to salvation. Many are really deists, that is they believe in a higher power or universal force that in the beginning wound up the clock and let nature take its course. The Bible shows this to be totally untrue as we are shown that God was active in setting the Jews free from the cruel Egyptian and Babylonian rulers and in leading them to the Promised Land. Also God in Jesus Christ sent His son to set us free from sin and reconcile us to God. You cannot be more active in human affairs than that!

In short we in the church have our struggles with sin and the church is there to help us in this endeavour. No one sin is more or less than another. We may have struggles with various sinful addictions such gambling, alcohol, drugs, adultery, homosexuality, gossip, lying, stealing, coveting, pornography etc. As committed Christians we will need help from the Holy Spirit working through the church to help overcome these sinful behaviours through prayer, counselling followed hopefully by eventual repentance. In short, the church should not condone any behaviour within the church considered sinful by God's word by those refusing to repent. However, they should try in Christian love to help that person overcome and repent from their sin addiction.

Sin is serious.

The basic command of God is to love God and our neighbour as ourselves (Mark 12:29–31). God emphasizes the seriousness of sin by saying that one should pluck out an eye or remove a limb if it causes one to sin (Matt 18:8, 9) stating that it would be better to enter heaven maimed than to end up in hell for all eternity. Of course Jesus was underlining the serious consequences of sin and not advocating self-mutilation as this really wouldn't be loving oneself as He commanded us to do (Mark12:29–31).

Nowadays we tend to justify our sinful behaviour by saying things like 'God made me this way' or 'My wife/husband doesn't really understand me and that is why I committed adultery' or 'You would have done these things if you had my dysfunctional childhood' or 'I am the product of evolution and blind chance so there really isn't any right or wrong.' It was no different in Adam's day when he made excuses and blamed God and his wife Eve for tempting him to sin (Gen 3:12).

The Bible says that we have all sinned and fallen short of God's glory, so in essence we all deserve an eternity in hell (Rom 3:23) and if we say that we have no sin, then we are calling God a liar (1 John1:10). And that is why God as Jesus died for us so that we could be reconciled to God, have the righteousness of Christ and spend eternity with our God. In fact there is no other way that we can be saved except through accepting Jesus Christ as our Lord and Saviour (John 3:16, Acts 16:31, John 14:6).

A good analogy of our limited goodness can be illustrated geographically. In Australia Perth is about 5,000 km west of Sydney. Let us just say that this 5,000 km represents a ruler of our righteousness. Hitler's good deeds might get him 10 km east of Perth, and Mother Theresa maybe 500 km east. But God's righteousness starts at Sydney and so we fall horribly short of God's requirements of being good enough to spend eternity with Him. Jesus through His death and resurrection gave us His righteousness or level of goodness (ie bridged this 5,000 km) so that we could spend an eternity with Him in Heaven.

As we give our hearts to God, His Holy Spirit works through us and changes us from within. As I look back on my life, I can see how God has changed me from within. As the Bible states we are saved by grace and not by works that any man may boast. Nonetheless, we are called to do the works that Christ would have us do (Eph 2:8–10). So to recap my good works aren't enough to justify me before God, only God's imparted goodness to me through the cross of Calvary. Grace is an unmerited gift from God to us, a reflection of His love for us. I love the words written by a minister called Todd White on grace where he says, 'Grace isn't a pass for us to sin but the power for us to live the way Jesus did, without sin.'

We live in a fallen world.

This world we live in on planet Earth is far from perfect with diseases, wars, natural disasters, financial stresses, uncertainties in life, criminal activity etc. When disasters strike, we tend to blame God even those who don't believe in God. These natural disasters are even called acts of God. The comments often made are how could a God of love allow these horrible events to happen. Therefore, they logically conclude that there is no God. My major disasters were loss of my son, my wife, and my divorce. However, I never blamed God for these events,

The Bible from the beginning explains clearly that it was man's sin that separated us from God resulting in death, suffering and this imperfect or fallen world (Gen 2:17; 3:16–19; Rom 5:12). Jesus came into this world to negate the effects of sin and out of love for us show us a sinless life and how we could live it in a more perfect and fruitful way. More importantly He came to redeem us from our sinful or sin-inherited natures and reconcile us back to God so that we could spend an eternity in His presence and in His kingdom. In other words, return us to the pre-fall Edenic existence. He did this by taking our sinful nature to the cross of Calvary 2,000 years ago thereby replacing our unrighteousness with His righteousness. All we had to do to inherit this eternal life was to admit our sinfulness, repent of it, ask God to forgive us and finally ask Him to be Our Lord and Saviour. As a sign of our acceptance by God, He fills us with His Holy Spirit. And when this happens, we know that we are now friends and children of God (1 John 3:2; John 15:15) In my experience after this event, one finds certain behaviours not acceptable anymore. For example, before conversion I used to get drunk at parties but after conversion lost the desire to do so. I became heightened to the inappropriateness of certain sinful behaviours. Christians believe that this is the influence of the Holy Spirit and God creating a new heart within us (Jer 24:7). That is why Jesus said we must be born again in order to inherit the kingdom of God (John 3:3).

Jesus often spoke of His heavenly kingdom and made comments that His kingdom is not of this world (John 18:36). He also spoke of preparing a place for us in His kingdom (John 14:2). In my heart, I know that this earthly life is only a temporary place and that I now have eternal life (Jesus Christ) living within in me right now (Gal 2:20). However, while I am in this flesh, I will always have a tendency to commit sin whilst in my resurrected body this will not be so (1 Cor 15:53, 54).

Unfortunately, there is a demonic spiritual world out there.

Paul the apostle alludes to this in his letter to the Ephesians when he says, 'For we do not wrestle against flesh and blood, but against principalities, against powers, against the rulers of darkness of this age, against spiritual hosts of wickedness in the heavenly places' (Eph 6:12).

Jesus and the apostles were often involved in releasing various people from the influence of demonic spirits. Hollywood would have us believe that the deliverance from the demonic was involved with much drama like screaming and frothing at the mouth. In reality the power of God is much greater than that of the satanic and deliverance in most situations is not so dramatic. When I came to the Lord, certain associates in and outside work were sometimes hurtful to me for no apparent reason at all, whereas before my conversion, they were quite friendly. I hadn't preached or even shared my conversion to Christ with them, which could be blamed for their hurtful behaviour toward me. Also when asked to preach at certain churches, opposition came from various people outside that church. In one instance my car barely got me to the church because of mechanical problems. After the service the car purred like a kitten. Eventually, the pressures from outside became so intense that I gave up preaching but not my church attendance. In hindsight I probably needed considerably more prayer support for my preaching activity.

To conclude I want to share another spiritual aspect related to this topic. Three months after I came to Christ, I received a call from a friend whose 21-year-old son had driven off in his car with a loaded 303 rifle with the expressed intent to shoot himself. This lad had been a heavy drinker and drug taker. I prayed for him and later in the day he returned home. I shared my conversion story and the love that Christ had for him. The words that I spoke to him just came effortlessly with a level of wisdom that I certainly wasn't aware of. In desperation I invited him to go to a Pentecostal Church meeting that evening.

At that stage I was a practicing Catholic and had never been to this Pentecostal service that was being held in a large school hall. The reason that I selected this place was because I had heard so-called miracles were happening there from other Christians. That evening there was about 600 attendees, and nobody knew us as we slithered quietly into the back seat of the hall. I desperately prayed to God under my breadth to help this disturbed young man, with no idea how God might help.

In the beginning of the service, there was some worshipful singing. The head minister then came to the podium and without any hesitation or prelims immediately described the young man sitting next to me to a tee, how drugs and alcohol had ruined his life and how he intended to end his life that night. You could have knocked me over with a feather. I had heard of word of knowledge being a gift of the Holy Spirit, but this time I actually experienced it. Anyway the minister asked this lad to raise his arm, which he did, the only arm raised in 600 people there. He then asked this lad to come to the front of the church for prayer.

I went to the front of this church with this lad for moral support. The minister brought him to Christ then laid hands on his head for the infilling of the Holy Spirit. The power of God was so strong that I was thrown to the floor in tears. The lad was shaking like a leaf besides being thirteen stone in weight. Anyway he was released and went on with Christ and is still alive forty years down the track. This story was truly a wonderful example of the spiritual warfare in this lad's life. How Satan tried to kill him and how Christ that Sunday night service had delivered him through the Holy Spirit inspired word of knowledge to the pastor. It was also a wonderful example of the power of Almighty God. All praise and glory to Him.

Another spiritual manifestation occurred about two to three months after my conversion back to Christ. I was going to do a weekend at the Nunyara retreat centre with 300–400 of the Catholic charismatic group. I was about to enrol for it but was informed that they had a full house. I was about to walk away disappointed when a man came to the desk saying that he had to work that weekend. I immediately jumped into his vacated place. On the Saturday we had teaching sessions mainly from Father Noel O'Brien.

On Sunday morning at about 1am, I could hear Fr Noel doing a deliverance prayer for a young man. I was on an upper bunk bed and decided to pray for this young man. Whilst praying, I had a vision of a brilliant white circular light about one foot in diameter on the opposite wall. I couldn't move or cry out and became oblivious of how long this vision lasted. Eventually it faded and I got down from my bunk to go to the toilet. I met a gentleman in the corridor who informed me that the chapel building at Nunyara was bathed in white light at 2 a.m. in the morning. At that time, I didn't know what all those visions meant.

The following Sunday morning we all went to mass. Fr Noel got up to preach but made an announcement that the Holy Spirit was going to take charge. So we all sat there like stunned mullets since a Catholic without a sermon is like a fish out of water. Slowly people got out of their chairs and gave short mini-testimonies like they once used to shoplift or swear a lot or lose their tempers easily. I was comfortable listening to these confessions, thinking that one day in the future when I was a mature Christian, that I would give my testimony. What happened next absolutely shocked me. I found myself standing in the front of a filled church with the microphone in my hand. To this day, how I got to the front of the church I don't know except it was the Holy Spirit having His way with me. Help, Lord! This was the only prayer my scared lips were capable of. My initial thought was that these folk were good Christians, far better than I, so how dare a sinner such as I preach to them. In sheer desperation with the mike in one hand, the other raised to heaven and both eyes closed, I gave my testimony about my infidelity, warts and all. It ran for about twenty to thirty minutes, but when I opened my eyes, the church was running to the altar rails for prayer. This was all very new to me, so I laid hands on the many folk and prayed for each one's needs. I had no idea that these folk were experiencing such traumas in their marriage and in the marriages of relatives and friends. I must have prayed for about one hour and was exhausted. When Denise my wife picked me up after, she asked me if I had been drinking because I had the look of one drunken soul. I assured her that I hadn't drunk but was drunk with the Holy Spirit. Similarly, in the Bible account, people thought that the apostles were drunk when they first preached the gospel (Acts 2:13–15). I hope my prayers helped all for whom I prayed for. One lady came up to me after and wanted me to pray for her husband who was about to move interstate and leave her and their three young kids for another woman the following weekend. A couple of days later, I shared my testimony with him, and he came back to his wife and kids. A few years later I bumped into them in a church that I had been invited to. They were all active in a music ministry in their church, praise be to God. One never knows how God will use us to His glory, but it is always exciting when He does. All praise to His wonderful name.

In my experience in life, there are several things that can push toward the demonic realm. The major one is unforgiveness and anger towards another person or persons, and I have covered this in the section on forgiveness. Taking drugs like alcohol, dope and others can have also a negative influence on our spiritual walk. Feeding our minds on pornography or entertaining sinful thoughts over a period of time is also spiritually dangerous. We are taught to be renewed in the spirits of our minds and think on positive and wholesome things (Rom 12:2; Eph 5:18–21). The other dangerous area is to dabble in the occult even innocently. These areas include Ouija board game, fortune tellers, sorcerers, witchcraft, consulting mediums and some New Age activities (Deut 18:10–12; Gal 5:19–21). I used to play Ouija and when I came to the Lord, I had to repent from this even though I had thought it was innocent fun like snakes and ladders. Repentance to the Lord is key to renewing your spiritual health. Satan was defeated at Calvary and therefore has zero power over us. So we don't have to ponder on him but to focus on Christ who has set us free (Gal 5:1).

Prayer is essential to a Christian life.

In the Bible we see that Jesus prayed often, sometimes all night, and taught His followers how to do likewise (Matt 6:9– 13, 7:7, 14:23; Mark 6:46, 14:32; Luke 6:12; John 17:9).

The apostles also encouraged us to pray often (1 Thess 5:17; James 5:13)

In my own life I have found praying to God to be a must, and I have seen God answer these prayers many times. It was prayer that saw my wife die without pain despite widespread metastases. Prayer saved that young man from suicide. It was through prayer that I gave my testimony. Prayer helped open up many research opportunities. Sometimes it feels that God hasn't heard my prayers as nothing seems to happen, but when I look back, I can see His hand on the problem. The only two words that I have ever heard from God is 'Trust Me,' usually when things seem impossible.

I need to mention two other events that I haven't touched on. One teenage lad I knew had gone off the rails and broken the law. He spent a couple of years in a teenage detention centre for his crimes. His family and I prayed for him for years. At about 16, he met and later married a lovely Christian girl. Initially, they had a rocky time, but about ten years later, he became a Bible-believing Christian, did a university theology and education degrees and then worked with kids having special needs. He and his wife became active Christians in their Baptist Church and ran several alpha courses. Fifty years on he is full on for the Lord.

In another situation, a friend of mine's mother who was 94 was dying. She was unable to speak and was close to death when we went to visit her with candles around her bed and photos of her loved ones. I sat by her bed thinking that she had had a full life, and it was time to go to be with the Lord. Out of the blue my friend asked me to pray for her mum. I wasn't expecting this request to pray. Anyway I gently touched her arm and asked God to bless her. Suddenly the Holy Spirit came on me, and I went into tearful prayer. The next moment she opened her eyes and started talking and did so with vigour for the next hour. Her son was overseas at this time. She died six months later, except this time all of her family were with her.

Is there life on other planets besides Earth?

One often hears or reads stories of people who have witnessed alien spacecraft in our atmosphere and how these crafts move at incredible speeds. Some folk have attested to the fact that they have been actually kidnapped and been operated on by aliens. These witnesses seem to be very sincere in what they have experienced about aliens. In fact many movies have been made on this subject.

Comments are frequently made that one would expect technologically advanced life, far in advance of our own, on other planets when one considers the huge size and age of our universe. In fact in the last 15 or so years, a complex of radio dishes, called SETI, have been listening to any messages that may come from outer space. These space ears have cost many millions of dollars. Nothing significantly has been heard up till now.

Dr Gary Parker from Creation Ministries and others have investigated this matter and have published a book and put out a DVD on this investigation entitled *Close Encounters of the Fourth Kind* (CMI 30-9-603). What they found was that these alien manifestations were never experienced by Christians who had had a born-again experience. Their conclusion confirmed that Satan the deceiver had duped people into believing certain alien events had occurred. In fact, this makes sense since Satan in the Bible has been described as coming across as an angel of light in order to deceive people (2 Cor11:14). In fact Jesus described Satan as a murderer and a liar from of old, it being part of the devil's nature (John 8:44). In short, therefore, these so-called alien manifestations are merely demonic deceptions

There is either a heaven or hell awaiting those post death.

To the atheist we just become manure post death. However, to many, death is just a transition from an earthly life to a spiritual one. That is our spirits and souls live on after death. That is, physically we die but spiritually we continue to live eternally. The Bible is quite clear in Hebrews 9:27 when it states that 'as it is appointed for man to die once but after that the judgment.' The judgement of the dead is also described in the Bible (Rev 20:11–15). So this puts an end to beliefs about reincarnation in certain religions and the atheist's manure story. Jesus taught about the torments in a place called hell (Matt 11: 23, 13:42, 25:41, 46) and the happiness of heaven where God Himself dwells (5:12, 13:43). There are many other Bible quotes on this subject.

As discussed, atheists deny the existence of God and a heaven and hell. I would rather believe God on this subject rather than the musings or speculations of man. Some believe that consideration of these subjects is fear mongering. However, it is important to face the truth. The Bible tells us to fear God is the beginning of wisdom (Psalm 111:10). This means to have reverence for God not to fear Him as one would a snake. I love and reverence God because He has blest us so much with grace, healing, joy, peace, salvation, creation, faith, the Holy Spirit and much more. All goodness and blessings flow from His throne. The main reward of heaven will be in being with Jesus Christ, my Lord and Saviour.

We often like to ponder on what heaven is like. Cartoonists depict it as a boring wispy place where we float around on clouds playing harps for eternity. This is certainly not scriptural. Paul who experienced heaven temporarily indicated it as a place too beautiful to describe (2 Cor 12: 1-4.! Cor 2:9). I believe it will be a place where we will continue to grow and learn. God describing heaven to us is like us trying to describe the outside world to a fetus. To the fetus the world is a watery dark place where one is fed through a tube. Telling the fetus about trains, nature, planes, computers, TVs, cars etc in the outside world would be beyond their comprehension. And so it is with us trying to fully understand heaven whilst on this earthly plane except that it is a place of great beauty, joy and peace and free from suffering.

It is important to work at one's spiritual life.

In life most people work at their physical health and this is important and encouraged biblically (3 John 2). My early days in a Catholic college encouraged sports probably more so than academic achievements in those days. In time, I developed a love of sports and ended up playing Australian rules football, athletics, tennis and squash. In my later years, at age 52, I took up a gentler sport in golf and have grown to love it. Playing sports helps keep me balanced and humble especially golf. My grandson Matthew placed a sticker on transparency on the rear window of my car that said, 'I work because I can't play golf.' I love his cheeky sense of humour.

Working at one's spiritual life appears to be more difficult, but I find it stimulating and enjoyable. Let me explain.

In my own life I pray 10–15 minutes every morning and often during the day. Whilst working, I offered up the day to the Lord and so my work became a prayer. I do a day-by-day study called Our Daily Bread, which comprises a short 10-minute study on different facets of life and how the Holy Scripture applies to them. During the day I keep up with news around the world on my PC that may have relevance to my Christian beliefs such as same-sex marriage, gender fluidity, euthanasia, abortion, violence against religious beliefs, freedom of religious beliefs, legalization of prostitution etc. I also do e-Bible (Got Ministries), which explains different aspects of our Christian beliefs from a biblical standpoint. I have also put considerable effort in understanding how certain Christian beliefs tie in with my understanding of God's creation from a scientific standpoint. I have attempted to discuss my understanding of this area in the rest of my book. All of these studies, taken together, have made me realize that the Bible is absolute truth. These studies take discipline, prayer and effort, but they really are a labour of love that I enjoy doing.

Is there a perfect Christian church I can join?

The answer to this is there isn't. The church consists of a group of sinners, so it is far from perfect. However, it should be emphasised from the beginning and from what I have already written that Christ is the only way to salvation for us sinners (John 14:6; Matt 7:13, 14; Rom 1:16).

There are, however, certain fundamental or foundational beliefs that the Christian church should follow. First is that we are all sinners in need of repentance from sin and that our salvation and redemption from sin is exclusively through the sacrificial death and resurrection of Jesus Christ at Calvary 2,000 years ago. Secondly, that Jesus Christ is truly God and truly man (1 John 4:2, 3), the second person of the divine Trinity together with God the Father and God the Holy Spirit (Col 2:9; 1 John 2:22, 23). Any belief that denies the deity of Christ and His human appearance on earth cannot be considered Christian (1 John 4:2). Thirdly, scripture is absolute truth inspired by the Holy Spirit and any addition or subtraction to this scripture must never be made even if the revelations come from so called angelic beings (Gal 1:6–9; Rev 22:18, 19; 2 Tim 3:16).

Within churches, there are peripheral differences like for example sprinkling versus full-immersion baptism, Saturday versus Sunday Sabbath, women's roles in the church, rules pertaining to divorce and remarriage, speaking in tongues etc. People can argue to all lengths as to who is correct in these topics. However, the foundational beliefs must be adhered for a church to be considered Christian.

There Are Two Basic Sciences – Experimental and Historical

Before I address the scientific aspects that impinge on my spiritual beliefs, two basic aspects of science need to be defined.

Experimental or operational science is that which is concerned with experiments. This type of science is repeatable in the lab and gives rise to discoveries of flat-screen TVs, smart phones, vaccinations, drugs for treating many diseases, putting man on the moon and so on. Experimental science is my field of endeavour.

Historical science is when one looks, for example, at a fossil in a rock and tries to figure how it got there and when. In other words, it is story telling because one cannot go back thousands or millions of years to see how it was formed. A secularist or atheist who believes life formed over many of millions of years would say the fossil in the rock was formed 3 million years ago. A creationist believing the Bible would say it was formed about 4 thousand years ago at the time of Noah's flood. So the interpretation would depend on whether one wore secular or biblical glasses. It is a non-repeatable experiment. Historical science is a bit like forensics where one tries to figure out how a crime was carried out in the past from the evidence around the crime scene.

As you read my discussion below, you will discern whether experimental or historical science relates to the discussion.

Naturalism versus Biblical Creationism

Most of what I have got to say in this section can be covered in greater detail by going to www.creation.com. Naturalism is a belief that everything in creation came about naturally without having to invoke God in this process. As such it is an atheistic or secular belief. It can also be called Darwinism as Charles Darwin is credited as being the promoter of this belief. It is also termed an evolutionary belief because everything we all see in our world evolved or came from basically a mixture of hydrogen and helium gases released from the big bang over 14 billions of years ago.

Biblical creation postulates that God created everything in our world and universe as described in the Bible. There are many scientists who believe in God our creator, including two of the greatest scientists ever namely Albert Einstein and Sir Isaac Newton. Young earth creationists (YECs) believe literally about the biblical account of creation in that it occurred about 6,000 years ago and not 14.6 billion years ago by chance. Creationists also believe that our Earth with all its life, land, water and sky, with its billions of galaxies was created in a six day period 6,000 years ago by Jesus Christ who is

God (John1:1–4: Col 1:15–17) as is described in the Bible (see Gen chapters 1 and 2).

There are Christians who try to accommodate the biblical creation story with the billions of years of evolutionary or Darwinian thinking. These Christians are therefore called theistic evolutionists.

My education from school to university was mainly biased toward the slow evolution concept or theory. I never questioned it or gave it much thought because it had no influence with my research endeavours. That is my research continued on quite happily whether I believed in evolution or not. I guess if asked at this time I would have said I was a believer in evolution because I trusted my professor's opinions and basically, like a sheep, went along with what most of my scientific colleagues believed.

When I became a Christian in 1977 there were pastors whom I met (not many) who ridiculed belief in evolution without any reason why given. I reasoned that they were not scientifically trained and therefore not competent to make such statements against evolution. My newfound faith in Christ had nothing to do with the evolutionary story, or so I thought.

However, with the passage of time certain parts of the Bible were hard to reconcile with my scientific background. For example, how could the whole population of the Earth come from one man, Adam and his wife Eve? There obviously would have been a lot of close relative intermarrying, and genetically, we know that would result in many deformed humans. Also how could Adam live to almost 1,000 years? My initial approach to these problems was that these were mysteries that only God had the answer to.

Clarification of these scientific difficulties came when Dr Carl Wieland of Creation Ministries International (CMI) spoke at our Thursday grand round at FMC in about early1990.

It was my ah-ha moment. The lecture theatre was packed with evolutionary thinkers, highly trained academic personnel. I was scared for Dr Wieland, thinking that he would be completely humiliated by this audience. That is, there would be something like a verbal tar and feathering of him after the talk.

Well, much to my surprise the audience was very quiet and civilized after the talk with barely a murmur. I approached Dr Wieland after the talk with the Adam and Eve questions re their old ages and close relative intermarrying. His answer was very simple and almost knocked me off my feet. Basically when they disobeyed God and were cast out of the Garden of Eden they were mutation free. Bingo, the scales fell from my eyes. What made it more poignant was that I had been working in the genetic mutation field for many years but was blinded to this simple answer to my question.

For those of you who may not be aware, mutations or changes to our DNA code is what causes us to age, develop cancers and other diseases of old age. DNA mutations also prevent us from close relative intermarrying because of the serious genetic diseases that would ensue from such unions. There is more on the mutational and other aspects of DNA later on.

Geologic Prodding

After this mind-blowing revelation, I started prodding other areas of science with what the Bible was saying. For example, from my geology classes many years before, I had been taught that rock strata were formed over many millions of years and that fossils in these strata were tens to hundreds of millions of years in age. This belief was reinforced by TV documentaries and articles published in the popular press. The story of Noah, the ark and God flooding the whole earth was presented as a myth in the Catholic and presumably many Protestant churches as well.

To cut a long story short, the geomorphology or land shapes of the earth are consistent with a worldwide flood. That is the huge depths of the rock strata and their distribution over hundreds of square miles are consistent with a worldwide flood. In fact, one PhD in geology I met, Dr Ron Neller, shared how he came to the Lord. He was an expert in sedimentation and geomorphology. Through his research around the Earth, he came to the conclusion that there had to be a worldwide flood to cause the geological features that he was studying. This conclusion didn't affect his long-held atheistic beliefs, but he thought that he would do a Bible study since his atheistic colleagues were labelling him a Christian creationist despite his protestations.

He is now a Christian and young earth creationist and works for Creation Ministries International (CMI) as one of their speakers.

The worldwide flood of Noah also showed that there were no intermediate fossils confirming the evolutionary story was a lie. In fact the fossil evidence was confirmatory of the sudden nature of the worldwide flood in that some fossils were found in the process of childbirth and also some in the middle of a meal. The geologic structures also confirmed that the rock layers were layered down rapidly within a period of days and not millions of years because of the lack of any weathering between layers and bioturbation (animals digging holes in the layers). Also the rock layers in many places were bent into synclines and anticlines (*n* and *u* shaped), confirming their formation from a watery soil deposition at about the same time. The presence of fossilized tree trunks bisecting a few rock layers also confirmed their deposition at about the same time. The fossilized tree fossil is called a polystrate because it cuts across a few rock layers, each layer supposedly formed over millions of years. There is more information on this subject in www. creation.com and in an excellent book called *How Noah's Flood Shaped Our Earth* by Michael J. Oard and John K. Reed, both highly trained scientists and specialists in this field.

Another argument used to confirm that rock layers were layered down in millions of years is supposedly confirmed by radioactive-dating methods. The radioactive elements used to date rocks have half-lives in the millions of years. That is the half-life is the time taken for the radioactive element, called the parent, to lose half of its radioactivity to give rise to the daughter element. The problem with this argument is that one never knows the starting amounts of the parent or daughter elements. There are also other factors such as the rate of radio decay has been shown to vary in the past, and there are considerations of contamination of the rocks with the elements concerned or alternately elution of these elements out of the rocks over time. For those confused by this discussion, radioactive decay can be likened to a bucket under a leaking tap. We may have gone on a short vacation and come back to find the bucket half full with water under a leaking, dripping tap. We can measure the rate of dripping at that time and the volume of each drip and calculate how long the tap has been dripping to half fill the bucket. However, we do not know if the tap initially was dripping faster (decay rate), how much water was in the bucket to begin with (daughter content), whether a sudden rain storm added to the water in the bucket (daughter contamination) or whether the dog next door had been drinking from this bucket (daughter isotope loss). As one can see there are a lot of ad hoc assumptions, one has to make in these calculations. Studies on recent igneous rocks of known age have shown erroneous results re the age of these rocks, out by millions of years.

One radioisotope is carbon-14 or C14 does however have some use because it has a short half-life of about 5,700 years. Because of its short half-life, it should not be detected in specimens after about 100–50,000 years. It has been detected in specimens reputedly millions to billions of years old such as fossils, coal, oil and diamonds thereby negating the so- called millions of years of age of our planet and confirming the thousands of years instead.

Dinosaurs fossils are believed to have died out 65 million of years ago. However, findings within the last 10 years have shown that T-rex bones have been found with red blood cells, DNA and stretchable proteins like collagen present (published in the journal Science by Dr Schweitzer about 1995). This evidence is consistent with dinosaurs being present a few hundred years ago on our planet as these biochemical are unstable and would certainly not survive a million years under any prescribed conditions. There has also been evidence of dinosaurs roaming the world in recent times (see www.creation.com).

DNA and Evolution

DNA is one area of science that I have had some experience in at the molecular level. In the evolutionary story, one basically goes from a microbe to a mouse to a monkey and finally a man. This process involves mutations and natural selection over many millions of years. DNA is essentially an information molecule that codes for our various characteristics like our height, body shape, eye, skin and hair colour and all our many other characteristics. In other words, it is the blue print that determines our various characteristics.

The DNA code is made up by four chemicals abbreviated A, C, G and T. These four chemicals are complex organic structures, called aromatic heterocyclic purine and pyrimidine bases. I won't draw these structures, but they can be found on Goggle or any biochemistry textbook for those interested. So basically our DNA code uses four letters of the alphabet to describe us. These letters must be in the correct order to make any sense of our code just like this book uses twenty-six letters of the English alphabet to describe what I am writing about. In other words, shaking these twenty-six letters in a box will not result in the information in this book.

Returning to the microbe to man story, one needs to go from a 3-million code for the microbe to a 3-billion code for man. For comparison, it is like going from a 100-page book (microbe) to 25 volumes each with 4, 000 pages (man). In biology that cannot happen in fact, it is impossible to go even from a 100- page book to a 101-page book containing new information in the extra page. As indicated in the fossil record, there is no evidence of intermediate life forms, for example like progressing from reptiles to birds. Yet scientists date rock stratum from the fossils present in that stratum. One can in the lab introduce a gene for making insulin into a microbe. And it will make insulin, but it will always be a microbe.

The process that causes a microbe to eventually become a man requires a combination of processes called mutations and natural selection. Mutation is simply a change to the code for example let say from GTACCTTGGTCA et cetera to GTACCT.GGTCA et cetera. The full stop means there has been a deletion or change (mutation) of the base T at position 7 from the left. This will result in the structure of nonsense, useless protein in most cases. There are many other types of mutations, and this field is complex suffice to say that mutations mostly arise in nonsense products within the cell. Evolutionists would say that the process of natural selection would get rid of these nonsense or deleterious mutations. However, natural selection generally is not sensitive enough to work at the molecular or mutation detection level.

Evolutionists and creationists do believe in natural selection (NS). NS is a culling process. For example, hairy dogs survive in cold climates, but hairless dogs do not. With your car you do not detect the slow rusting occurring under the paintwork. This is akin to natural selection not detecting mutations. However, when you detect the rust as it shows through the paint you then become like natural selection and either sell the car or remove the rust and repaint it. By contrast, in biology once a mutation has set in you cannot reverse it.

It has been estimated by geneticists that each generation passes on about 100 new mutations to the next generation. For those wanting more information, this has been described in a book by Dr John Sanford entitled *Genetic Entropy and the Mystery of the Genome*. Evolutionists believe that us (Homo sapiens) have been here for about 200,000 years. Geneticists have calculated that, if this was so, we should have become extinct thousands of years ago because of the mutational load accrued over that span of years. So it is a problem for those believing in evolution. It should be emphasized that creationists do believe in change within a kind but not change giving rise to a new kind. For example, we believe that there are many varieties of dogs from chihuahuas to Great Danes. But dogs will never become cats or horses. We believe the original dog kind were two wolflike animals that contained the genetic information that allowed all the dog kinds we now see today. In a similar vein, Adam and Eve had the genetic information to give all the variety that we now see in humans today. Evolutionists try to convince us that there were once ape-like creatures that eventually became us.

But that has since been debunked.

Going back to the microbe mentioned earlier, it is difficult to conceive how this so-called simple organism came about. How 3 million base pairs of the A, C, G, T alphabet could arrange itself in the right order for 3 million bases to form the first microbe. Not only does one need the bases to be in the right order, but one also requires thousands of proteins, over twenty transfer RNAs and a sophisticated membrane system both outside and inside the cell, the ability to energize itself through ATP production et cetera. In short having the right DNA sequence is just one factor in cell's autonomy to life. There are thousands of other factors to consider in the complex working of a cell. And even if we put all these factors together in a test tube, they would never arrange themselves into a living cell. In experiments we often pull the cell apart using special lab techniques. Once the cell has been so treated, despite having all the right ingredients for life, it will never reform itself under any favourable laboratory conditions.

According to the evolutionary thinking, life itself suddenly appeared in the Cambrian rock stratum hundreds of millions of years ago. However, one finds in reality many life forms in this stratum not just microbes. If evolution did happen, then one would expect to see more complex life forms in more recent rock strata. So the formations of the first life form, even though only a 3-million base pair microbe is difficult to explain. Some scientists have dodged this question saying that aliens seeded the first life forms. And even if that did happen, it is impossible to go from microbes to man.

As mentioned earlier, humans were mutation free when they left Eden. This explained their long lives and the ability to marry close relatives. However, as the mutation level rose, we find that the life spans decrease accordingly (compare Gen 5:3- 32 with Gen 11:10-25). This accords with the clinical research findings thast those with a high mutation rate die earlier in life, for example patients with Progeria. Later on in life, when the mutation rate had increased to a certain level, God forbade close relative intermarrying (Lev 18:6-17).

Evolution and Cosmology

This is an area that I am not totally familiar with. One of the major problems in this area is trying to get one's mind around a young 6,000-year-old universe and a universe that is believed to be about 14 billion light-years in diameter. In other words, how can light from galaxies 14 billion light-years away be seen in an Earth supposedly created 6,000 years ago?

According to Einstein, time is relative. In other words, time can vary. It is well known, for example, that gravity slows up time. If one considers the early universe where all the galaxies were closer together then the gravitational effects would have been very much greater. The Bible says that God stretched out the universe like a curtain (Psalm 104:2), and this would have increased the time clocks in the outer galaxies compared to the inner or central galaxies because the gravitational effects would have been much less in the outer galaxies with the stretching out. It is calculated that the time clocks in the outer galaxies would have been over a trillion times faster than the inner galaxies. And our Earth is part of the inner galaxies if not the central galaxy. This is one very plausible theory to explain the time travel phenomenon of light coming from distant galaxies. There are others, but they are too complex for me to get my mind around.

The other area of this field that I find difficult to understand is how could trillions of stars like our sun come into existence as a result of a big bang? Cosmologists agree that the initial elements released in the big bang were mainly hydrogen and lesser amounts of helium gas. To my mind these gasses would have dissipated, but somehow they formed the many stars and about 100 elements that we have in the universe today. To dodge this problem, evolutionists propose that there is dark matter and dark energy that is helping make our many galaxies and elements. However, despite much research, nobody has been able to find either dark energy or dark matter. So over 95 per cent of our universe is undetectable except in math's formulae. The first law of thermodynamics states that energy or matter cannot be created or destroyed but may be changed one to another. For example, a block of wood cannot appear or disappear from thin air. It may however be burned to give light, heat and ash residual energy. In other words, the big bang goes against this first law in that we get something out of nothing. In short the whole universe and the laws governing it scream out the need for a creator. And the Bible tells us that creator is Jesus Christ (John 1, 1–3; Col 1:16). To me everything in nature from the smallest cell to the vast universe shouts out the existence of a creator in this process.

I have shared my creation beliefs with Christians from time to time. Most times I have received a semi-hostile response where people say things like God didn't really create in six days but did so in six eras of time. My response is then why did God in Exodus 20 state that we should work six days and rest on the seventh just like He did when creating our universe and world. If we disregard biblical creation, we are faced with the option that creation was a slow evolutionary process resulting in human appearance about 200,000 years ago. This then leads to the proposal that there never was an Adam and Eve and therefore no original sin. This in turn means Jesus Christ did not need to come to Earth, die and redeem us from our sinful nature. This logic is used by atheists and makes perfect sense, except there was an Adam and Eve, and I have already discussed their reality. Also experts in genetics know the mutation rates in humans and have reasoned that if we are 200,000 years on Earth, then we should have died out many times over.

Bad Press on Religion

Religions in general have received bad reviews from the public, much of it well deserved. In recent times we hear of paedophilia by priests within churches and public institutions. In the past there have been wars in the name of religion/God such as the crusades, inquisitions, terrorism, witch-hunts and the like. This religious barbarism, which resulted in the deaths of thousands, went directly against the word of God where we are exhorted to love and pray for our enemies (Matt 5:44). I am sure that these religious activities done in the name of God saddened God greatly.

Perhaps the greatest falling away from God occurred in the twentieth century and was probably aided by a strong belief in human reasoning separate from the Bible and to a large extent, in evolutionary beliefs where God is non-existents and therefore irrelevant. In this twentieth century with reduced belief in God we saw the growth of communism, Nazism, fascism, racism. These new beliefs resulted in wars and the slaughter of over 200 million people. Also in the middle of the twentieth century, where religious beliefs are considered old hat and downgraded, that we see a dramatic increase in the crime rate. One also experiences acceptance of abortion, euthanasia, prostitution, same-sex marriage and surgically driven gender fluidity and reassignment as normal. To summarize, Christianity has had a positive effect in people's lives. Where one downplays the importance of Christian beliefs in a society then that society pays a heavy price, and history bears this out.

Conclusion

Looking back over my life, I can see God's hand in directing my paths. My life's journey is not meant to be a bragging session and I am no better than anyone else. I am sure God works just as significantly in the lives of many others. This book is just to describe my own journey. Coming to God in my fortieth year was a significant turning point to the better for me. Without a commitment to God, I doubt if I would have lived long and achieved as much in my life and job. My initial Catholic upbringing had a significant role in setting my spiritual foundation for my conversion later on, even though as a Catholic I didn't experience a personal relationship with Jesus. His reality only came when I said the sinner's prayer later on which eventually led my receiving the Holy Spirit's baptism. It must be emphasized that these spiritual events were God instigated when I opened my heart to Him. It didn't come from a detailed theological study of the Lord or years of study.

My conversion reminded me of Christ's conversation with Nicodemus a Pharisee priest well versed in the Old Testament (see John 3). At night, he snuck in to talk with Jesus when no one was looking. Jesus cut to the quick and told him that he needed to be born again to enter the kingdom of heaven, no heavy theological discussion here. Jesus didn't mince words but always went to what was really important. Nicodemus was confused thinking how could he get back to his mother's womb to be born again. Jesus explained that those born bodily couldn't enter the His kingdom, but only those born of the Holy Spirit. And that could only happen by admitting our sinful nature before God and asking not only God's forgiveness for our sins but also to be Our Lord and Saviour. Once I did that I had a born-again experience although at the time all I felt was a deep peace come into my tormented soul. Through the passage of time, I came to understand this born-again experience and why we all need to go through it because we have all sinned and fallen short of God's level of righteousness and thereby need to be reconciled to Him (Rom 3:23, 24).

King Solomon asked God for wisdom and became well known for this. In fact kings and queens from far-off lands came to meet this remarkable man. In addition to his God- given wisdom, he had been blessed with great wealth and had over 300 wives and concubines. He spurned this material wealth and came to realize that God alone could satisfy our deepest needs. C S Lewis a professor of English and author of many well-known books came to a similar conclusion and basically stated that God couldn't bless us or make us content without Christ in our lives. Lewis was, for a considerable time, an atheist and later converted to Christianity. St Augustine stated that a soul could only be at peace till it rested in Christ. And he was a guy who once tasted of many humanistic lifestyles.

There are many decisions that we make in life, some good and some bad. The most significant decision that we can make in life will have eternal consequences, and this is the decision to follow God in Christ or not. A scientist called Blasé Pascal proposed that we follow God as an insurance policy for getting to heaven (called Pascal's wager). However, this is not acceptable because God wants us to have a deep personal relationship with Him (Matt 22:37). In fact, God said that He would vomit such a lukewarm relationship from His mouth (Rev 3:16). In other words, there is no sitting on the fence with Our Lord. We are either 100 per cent for or against Him. In fact following the Lord on Earth not only has consequences for our eternal happiness but also leads to a more fruitful and joyous life on Earth.

The take-home message of this book is eternal salvation is ours simply if we acknowledge our sinfulness before God, ask Him to forgive our sins and become Our Lord and Saviour. In addition, your life in this physical world will take on significant meaning when you make Jesus Christ your Lord and Saviour as His Holy Spirit infills, directs and comforts your life. It won't be a smooth life. As Jesus once said to take up your cross and follow Him (Matt 10:38). However, you will experience the joy, peace and leading of the indwelling Holy Spirit regardless of your circumstances. As the hymn says, the things of this world will grow dim in the presence of His glory and grace. Also don't be discouraged by some Christians that you may meet along the way. They, like us, are human and need prayer to follow God faithfully. Always keep your trust and focus on Christ and not on fallible human kind. We are always connected to God as we live in love, charity and forgiveness. God will always work to restore our relationship with Him. As the Bible says nothing can ever separate us from the love of Almighty God (Rom 8:38, 39).

In our current world, we are tempted by materialism, sex, and success, but as others have experienced already, this only leads to emptiness within and that only God can fill this emptiness. This has also been my experience and those of others so be warned. I hope that this book has been a blessing to you. With God's richest blessings, love Ainsley.

My commentaries on certain biblically related subjects may also be Googled on ainsley chalmers on www.ebible.com

Book published by author
  A H Chalmers, 'The Physical Body, the Spiritual Body,' 94 pages Balboa Press 2016. ISBN:978-1-5043-0496-2 (sc). ISBN: 978-5043-0497

Publications – A H Chalmers

1. A H Chalmers, C C J Culvenor and L W Smith, 'Characterization of pyrrolizidine alkaloids by gas, thin layer and paper chromatography,' *J. Chromatography* 20: 270–277 (1965).
2. A H Chalmers, P R Knight and M R Atkinson, 'Conversion of azathioprine into mercaptopurine and mercaptoimidazole derivatives *in vitro* and during immunosuppressive therapy,' *Aust. J. Exp. Biol. Med. Sci.* 45: 681–691 (1967).
3. R P Rao and A H Chalmers, 'Indole derivatives: Part 1 – Preparation and reaction of some alkylaminoethylindoles and related compounds.' *Indian J. Chem.* 6: 336–340 (1968).
4. A H Chalmers, P R Knight and M R Atkinson, '6-Thiopurines as substrates and inhibitors of purine oxidases: A pathway for the conversion of azathioprine into 6-thiouric acid without release of 6-mercaptopurine,' *Aust. J. Exp. Biol. Med. Sci.* 47: 263–273 (1969).
5. A H Chalmers, M R Atkinson, L A Burgoyne and A W Murray, 'Evidence for the utilisation of preformed purines by rat lymphoid tissues: Implications to immunosuppression by 6-thiopurine analogues,' *Aust. J. Exp. Biol. Med. Sci.* 46: 177–183 (1971).
6. A H Chalmers, T Gotjamanos, M Mohanrao, P R Knight and M R Atkinson, 'The immunosuppressive activity of 9-butylazathioprine

used alone and in combination with azathioprine,' *J. Surgical Res.* 11: 284–288 (1971).

7. G R Donaldson, A H Chalmers, M R Atkinson and A W Murray, 'The effect of 9-butyl-6 thiopurines on enzymes which metabolise purines,' *Aust. J. Exp. Biol. Med. Sci.* 49: 121–124 (1971).

8. A H Chalmers, T Burdorf and A W Murray, 'Immunosuppression by 9-alkyl-6-thiopurines,' *Biochem. Pharmacol.* 21: 2662–2664 (1972).

9. A H Chalmers, L A Burgoyne and A W Murray, 'Antineoplastic and immunosuppressive drugs 1: Biochemical and clinical pharmacological considerations,' *Drugs* 3: 227–253 (1972).

10. A H Chalmers, 'The mechanism of formation of 5-mercapto-1-methyl-4-nitrimidazole during immunosuppressive therapy with azathioprine,' *Biochem. Pharmacol.* 23: 1891–1901 (1974).

11. A H Chalmers, A W Murray and P Verakalasa, 'The synthesis and immunoenchancing activity of 3-butylazathioprine,' *Adv. Exp. Biol. Med.* 41: 699–704 (1974).

12. A H Chalmers, 'A spectrophotometric method for the estimation of urinary azathioprine, 6-mercaptopurine and 6-thiouric acid,' *Biochem. Med.* 12: 234–241 (1975).

13. D W Thomas, B Hannett, A H Chalmers, A M Rofe, J. Edwards and R G Edwards, 'Oxalate excretion during carbohydrate infusion,' *Int. J. Vit. Nutrition Res. Suppl.* 15: 181–192 (1976).

14. D W Thomas, B. Hannett, A H Chalmers, A M Rofe, J B Edwards and R G Edwards, 'The formation of oxalate in pyridoxine deficient rats following intravenous xylitol infusion,' *J. Nutrition* 7: 458–465 (1977).

15. A H Chalmers, M Lavin, S Atisoontornkul, J Mansbridge and C Kidson, 'Resistance of human melanoma cells to ultraviolet radiation,' *Cancer Res.* 36: 1930–1934 (1976).

16. M Lavin, A H Chalmers, and C Kidson, 'DNA repair and U.V. resistance in human melanoma,' *Molecular Mechanisms for Repair of DNA* 5: 817–819 (1975).

17. M F Lavin, G M Willett, A H Chalmers and C Kidson, 'DNA

replication and repair n a human melanoma cell line resistant to ultra-violet radiation,' *Int. J. Radiation Biol.* 31: 101–111 (1977).

18. C Farrington and A H Chalmers, 'The dilution effect of CPK,' *Clin.Chim. Acta* 73: 217–219 (1976).

19. A M Rofe, A H Chalmers and J B Edwards, 'C14- Oxalate synthesis from U-C14 glyoxalate and 1-C14 glycolate in isolated rat hepatocytes,' *Biochem. Med.* 16: 277–283 (1977).

20. M G Tingay, A H Ilsey, R J Willis, M J Thompson, A H Chalmers and M J Cousins, 'Gas identity hazards and major contamination of the medical gas system of a new hospital,' *Anaesth. Intens. Care* 6: 202–209 (1978).

21. C J Farrington and A H Chalmers, 'The estimation of urinary oxalate by gas chromatography and its comparison with a colorimetric method,' *Clin. Chem.* 25: 1993–1996 (1979).

22. C J Farrington, M Liddy and A H Chalmers, 'A simplified sensitive method for the analysis of renal calculi,' *Am. J. Clin. Pathol.* 73: 96–99 (1980).

23. C J Farrington, M L Liddy and A H Chalmers, 'Un metodo simply sensible para el analysis de calculos renales,' *Analisis Clinicos* 4: 1–4 (1981).

24. M N Berry, R D Mazzachi and A H Chalmers, 'A radio-enzymatic ultramicro method applicable to the measurement of a wide range of metabolites,' *Analyt. Biochem.* 118: 344–352 (1981).

25. A H Chalmers, 'A simplified method for estimating 5-phosphoribosyl 1-pyrophosphate in mouse liver and spleen,' *Aust. J. Exp. Biol. Med. Sci.* 62: 281–290 (1984).

26. A H Chalmers, M Mohan Rao and V R Marshall, 'The effect of purine and pyrimidine bases on splenic plaque forming cells and cellular immunity,' *Aust. J. Exp. Med. Sci.* 62: 269–279 (1984).

27. A H Chalmers and D M Cowley, 'Urinary oxalate by rate analysis compared with gas chromatographic and centrifugal analyser methods,' *Clin. Chem.* 30: 1891– 1892 (1984).

28. D M Cowley, A H Chalmers, B M Mottram, M Peters and T J Sinton, 'Collection of 24-hour urine specimens,' *The Clin. Biochem. Newsletter* 74: 16–17 (1984).

29. D M Cowley, B A Nagle, A H Chalmers and T J Sinton, 'Collection of specimens for plasma ammonia analysis: Effects of platelets,' *Clin. Chem.* 31: 332–333 (1985).
30. B F Hughes, K A Rye, L B Pickford, G J Barritt and A H Chalmers, 'A transient increase in diacylglycerols is associated with the action of vasopressin on hepatocytes,' *Biochem. J.* 222: 535–540 (1984).
31. A H Chalmers and J Lark, 'Urinary ascorbate measurement by rate analysis in a centrifugal analyser,' *Clin. Chem.* 31: 353–354 (1985).
32. A H Chalmers, T Rotstein, M Mohan Rao, V R Marshall and M Coleman, 'Studies on the mechanism of immunosuppression with adenine,' *Int. J. Immunopharmacol.* 7: 433–442 (1985).
33. C Holt, D M Cowley and A H Chalmers, 'The rapid estimation of citrate using a centrifugal analyser,' *Clin. Chem.* 31: 779–780 (1985).
34. A H Chalmers, D M Cowley and B C McWhinney, 'Stability of ascorbate in urine: relevance to analyses for ascorbate and oxalate,' *Clin. Chem.* 31: 1703–1705 (1985).
35. A H Chalmers and D M Cowley, 'Stabilization of the reaction mixture used in urinary citrate estimations,' *Clin. Chem.* 31: 1579 (1985).
36. B C McWhinney, D M Cowley and A H Chalmers, 'An automated method for measuring plasma citrate without protein precipitation,' *Clin. Chem.* 31: 1578–1579 (1985).
37. J A Renouf, Y H Thong and A H Chalmers, 'A rapid simple method for the microestimation of purine nucleoside phosphorylase activity in peripheral blood lymphocytes,' *Clinica Chimica Acta* 151: 311–316 (1985).
38. A H Chalmers, D M Cowley and J M Brown, 'A possible etiological role for ascorbate in calculi formation,' *Clin. Chem.* 32: 333–336 (1986).
39. A H Chalmers and B C McWhinney, 'Two spectrophotometric methods compared for measuring low concentrations of ascorbate in plasma and urine,' *Clin. Chem.* 32: 1412 (1986).

40. B C McWhinney, D M Cowley and A H Chalmers, 'A simplified liquid chromatographic method for measuring urinary oxalate,' *J. Chromatog. Biomed. Appl.* 383: 137– 141 (1986).
41. J M Brown, A H Chalmers, D M Cowley and B C McWhinney, 'Enteric hyperoxaluria and urolithiasis,' *New Engl. J. Med.* 315: 970–971 (1986).
42. J M Brown, G Stratmann, D M Cowley, B M Mottram and A H Chalmers, 'The variability and dietary dependence of urinary oxalate excretion in recurrent calcium stone formers,' *Ann. Clin. Biochem.* 24: 385–390 (1987).
43. D M Cowley, B.C. McWhinney, J.M. Brown and A H Chalmers, 'Chemical parameters important for calcium nephrolithiasis: evidence of impaired hydroxycarboxylic acid absorption causing hyperoxaluria,' *Clin. Chem.* 33: 243–247 (1987).
44. B C McWhinney, S L Nagel, D M Cowley, J M Brown and A H Chalmers, 'Two carbon oxalogenesis compared in recurrent calcium oxalate stone formers and normal subjects,' *Clin. Chem.* 33: 1118–1120 (1987).
45. J A Renouf, Y H Thong and A H Chalmers, 'Activities of purine metabolizing enzymes in lymphocytes of neonates and young children: correlates with immune function,' *Immunol. Letters* 15: 161–166 (1987).
46. B S Teh, W K Seow, A H Chalmers, S Playford, B Iannoni and Y H Thong, 'Inhibition of histamine release from rat mast cells by the plant alkaloid tetrandrine,' *Int. Arch. Allergy Appl. Immunol.* 86: 220–224 (1988).
47. A H Chalmers, D M Cowley, J M Brown and B C McWhinney, 'More on citric acid and calcium nephrolithiasis,' *Clin. Chem.* 34: 793 (1988).
48. W K Seow, A Ferrante, D B H Goh, A H Chalmers, Li Si-Ying, and Y H Thong, '*In vitro* immunosuppressive properties of the plan alkaloid tetrandrine,' *Int. Arch. Allergy Appl. Immunol.* 85: 410–415 (1988).
49. D M Cowley, T M Brown, B C McWhinney, A H Chalmers, 'Hydroxy carboxylate malabsorption and calcium exalted

nephrolithiasis in urolithiasis' (ed. by R A L Sutton, E C Cameron, V Walker, B Robertson, and C C Pak) *Plenum Press New York* p. 481–3 (1989).

50. D M Cowley, B C McWhinney, J M Brown and A H Chalmers, 'Effect of citrate on the urinary excretion of calcium and oxalate: relevance to calcium oxalate nephrolithiasis,' *Clin. Chem.* 35: 23–28 (1989).

51. D M Cowley, J M Brown, B C McWhinney and A C Chalmers, 'Hydroxycarboxylate malabsorption and calcium oxalate nephrolithiasis' *Urol. Res.* 16: 179 (1989).

52. J A Renouf, A Wood, I H Frazer, Y H Thong and A H Chalmers, 'Depressed activities of purine enzymes in lymphocytes of patients infected with human immunodeficiency virus,' *Clin. Chem.* 35: 1478–1481 (1989).

53. E A DeLeacy, D M Cowley, J M Brown, B C McWhinney and A H Chalmers, 'Effect of oral citrate on urinary calcium excretion after a load of oral calcium phosphate,' *Clin. Chem.* 35: 1541 (1989).

54. B Ioannoni, A H Chalmers, W K Seow, J G McCormack and Y H Thong, 'Tetrandrine and transmembrane signal transduction: Effect on phophoinositide metabolism, calcium flux and protein kinase C translocation in human lymphocytes,' *Int. Arch. Allergy Appl. Immunol.* 89: 349–54 (1989).

55. A H Chalmers, C Hare, 'A semi-automated method for 5'-ectonucleotidase measurement in lymphocytes,' *Immunol. Cell Biol.* 68: 75–79 (1990).

56. A H Chalmers, C Hare, G Woolley and I H Frazer, 'Lymphocyte ectoenzyme activity compared in healthy persons and patients seropositive to or at high risk of HIV infection,' *Immunol. Cell Biol.* 68: 81–85 (1990).

57. A H Chalmers, 'Simple, sensitive measurement of carbon monoxide in plasma,' *Clin. Chem.* 37: 1442–1445 (1991).

58. A H Chalmers and L E Snell, 'Estimation of plasma and urinary haemoglobin by a rate spectrophotometric method,' *Clin. Chem.* 39: 1679–1682 (1993).

59. A H Chalmers, 'Fresh substrate essential for the transketolase reaction,' *Clin. Chem.* 39: 1347–1348 (1993).
60. B Ioannoni and A H Chalmers, 'Increased calcium absorption in nephrolithiasis explained by uptake studies in ileal brush border membrane vesicles,' *Biochem. Med. Met. Biol.* 51: 99–104 (1994).
61. R Seshadri, D J Horsfall, F A Firgaira, K McCaul, V Setlur, A H Chalmers, R Yeo, D Ingram, H Dawkins, and R Hahnel, 'The relative prognostic significance of total Cathepsin-D and HER-2/neu oncogene amplification in breast cancer,' *Int. J. Cancer* 56: 61–65 (1994).
62. A H Chalmers, 'Ascorbic acid bioavailability in foods and supplements,' *Nutrition Review* 52/3: 110 (1994).
63. J E Buttery, B R Chamberlain and A H Chalmers, 'Fresh versus frozen substrate for transketolase assay,' *Clin. Chem.* 40: 1786–1787 (1994).
64. A H Chalmers, 'Ascorbate acid overdosing: A risk factor for calcium oxalate nephrolithiasis,' *J. Urol.* 152: 171 (1994).
65. A H Chalmers and M J Peake, 'Underestimation of serum iron using automated methods,' *Clin. Chem.* Clinical Chem 41: 1199–1200 (1995).
66. J S Blake-Mortimer, A H Winefield and A H Chalmers, 'The relationship between 5'-ectonucleotidase and psychological stress,' *International J Stress Management 3: 189–207 (1996)*.
67. M Galluccio and A H Chalmers, 'Assay of serum folate is not required for assessment of tissue folate stores,' Aust J Med Sci., 17: 148–149 (1996).
68. J S Blake-Mortimer, A H Winefield, A H Chalmers, 'Evidence for free radical-mediated reduction of lymphocytic 5'-ectonucleotidase during stress,' International J Stress Management 5: 57-75 (1998).
69. A H Chalmers, M Kiley. 'Detection of xanthochromia in cerebrospinal fluid,' Clin Chem 44: 1740–2 (1998).
70. J S Blake-Mortimer, A H Winefield, A H Chalmers, 'The effect of depression in an animal model on 5'-ectonucleotidase, antibody production and tissue ascorbate stores,' *The Journal General Psychology* 125: 129–146 (1998).

71. A H Chalmers, 'Smoking and Oxidative stress,' Am J Clin Nutrit 69:572 (1999).
72. 72. A H Chalmers, '5'-Ectonucleotidase in chronic lymphocytic leukemia,' Clin Biochem 32: 91 (1999).
73. 73. P Bierer, A W Holt, A D Berstein, J L Plummer, A H Chalmers, 'Haemolysis associated with continuous venovenous renal replacement circuits,' Anaesth. Intensive Care 26: 272-5 (1998).
74. A H Chalmers, J S Blake-Mortimer, A H Winefield, 'Lymphocytic 5'-ectonucleotidase, an indicator of oxidative stress in humans,' *Redox Report* 5: 89–91 (2000)
75. A H Chalmers, 'CSF Xanthochromia simplified,' Clin Chem. 47: 147–8 (2001).
76. J R Hapuarachchi, A H Winefield, A H Chalmers, J S Blake-Mortimer, 'The impact of work-stress on health and wellbeing in university staff: changes in clinically relevant metabolites with psychological strain,' Int J Behavioural Med 9: 103–4 (2002).
77. A H Chalmers, J S Blake-Mortimer, A H Winefield, 'The prooxidant state and psychologic stress. Environmental Health Perspectives' 111: A16–A17 (2003).
78. J R Hapuarachchi, A H Chalmers, A H Winefield, J S Blake-Mortimer, 'Changes in clinically relevant metabolites with psychological stress parameters. Behavioral Medicine' 29: 52–59 (2003).
79. A H Chalmers. 'Comments on the proposed national guidelines for analysis of cerebrospinal fluid for bilirubin in suspected subarachnoid haemorrhage. Produced by a working group of UK NEQAS for Immunochemistry and in conjunction with the National Audit Group of the Association of Clinical Biochemists,' Clin Biochem Rev 24: 132–4 (2003) -9 (e).

Abstracts of Scientific Meetings – A H Chalmers

1. A H Chalmers, P R Knight and M R Atkinson (1968). Aust. and New Zealand J. of Surgery, 37, 319. 'Metabolism of azathioprine.'
2. A H Chalmers, P R Knight and M R Atkinson (1969). Aust. and New Zealand J. of Surgery, 38, 179. 'Hydroxylation of azathioprine 6-mercaptopurine.'
3. A H Chalmers, P R Knight and M R Atkinson (1968). Proc. Aust. Biochem. Soc., 1, 15. 'Oxidative metabolism of azathioprine in immuno-suppressive therapy.'
4. A H Chalmers, A W Murray, T Gotjamanos, P R Knight and M R Atkinson (1969). Proc. Aust. Biochem. Soc., 2, 38. 'Studies of the immunosuppressive activity of 6-thiopurines and their butylated derivatives.'
5. T Gotjamanos, P G Gill and A H Chalmers (1970). Aust. and New Zealand J. of Surgery, 39, 316. 'Changes in liver cell morphology and function induced by cortisol, azathioprine and anti-lymphocyte serum.'
6. A H Chalmers, A W Murray, L A Burgoyne and M R Atkinson (1970). Aust. and New Zealand J. of Surgery, 40, 314. 'Inhibition

of DNA synthesis in rat lymphoid tissue by immunosuppressive 6-thiopurine analogues.'
7. A.H. Chalmers, A.W. Murray and T. Burdorf (1971). Proc. Soc. Aust. Med. Res., 2, 445. Immunosuppressive properties of 9-alkyl-6-thiopurines.
8. V R Marshall, D C Hoffman and A H Chalmers (1972). Aust. and New Zealand J. of Surgery, 42, 100. 'The preparation and assay of carcinoembryonic antigen CEA.'
9. A H Chalmers and C S Kidson (1973). Proc. Aust. Biochem. Soc., 6, 53, 'The repair of DNA in human melanoma cells resistant to ultra-violet radiation.'
10. A M Rofe, A H Chalmers and J B Edwards (1975), Clinical and Experimental Pharmacol., 2, 70. 'Carbohydrate metabolism in isolated rat levier cells.'
11. D W Thomas, B Hannett, A H Chalmers, J B Edards and R G Edwards (1975). Clin. Chem., 21, 957. 'Intra- and extracellular changes induced by infusions of hypertonic carbohydrate solutions.'
12. A H Chalmers (1977), The Proc. Clin. Oncolog. Soc. Aust., p. 27. 'The Effect of purine and pyrimidine bases on the immune response.'
13. A H Chalmers and T Rotstein (1980), The Proc. Clin. Oncolog. Soc. Aust., p. 49. 'Investigation into the mechanism of immunosuppression shown by adenine and adenosine.'
14. A H Chalmers and T Rotstein (1981), The Proc. Aust. Biochem. Soc., 14, 90. 'The role of 5-phosphoribosyl-1- pyrophosphate in the immune response.'
15. A H Chalmers, D M Cowley, B C McWhinney and J M Brown (1986), *The Clinical Biochemist Reviews*, 7, 80, 'The role of imparied hydroxycarboxylic acid absorption in recurrent calcium nephrolithiasis,' presented at the National AACB meeting Hobart, Tasmania, October, 1986.
16. J A Renouf, Y H Thong and A H Chalmers (1986), 'Activities of purine enzymes in the cord blood lymphocytes of premature infants,' poster presented at the first IUIS conference of Clinical Immunology, Toronto, Canada, July 1986.

17. J A Renouf, Y H Thong and A H Chalmers (1986), 'Micromethods for the estimation of purine enzyme activities in lymphocytes from neonates.' Poster presented at the Third International Congress of Paediatric Laboratory Medicine, Bristol, England, June 1986.
18. A H Chalmers, J A Renouf, and Y H Thong (1986), 'Depressed purine enzyme activities in cord blood lymphocytes premature and post-term infants,' Presented at the Australia Perinatal Society 4th Congress, Brisbane, September 1986.
19. A H Chalmers, J A Renouf, I Frazer, F M Crapper and Y H Thong (1987), 'Depressed purine enzyme activities in lymphocytes from patients with HIV infection,' Ann. Clin. Biochem 24 (Suppl. 2), 110. Poster presented at the 13th International Congress of Clinical Chemistry. The Hague, Holland, June/July 1987. Poster awarded first prize at the meeting.
20. B S Teh, A H Chalmers, W K Seow, B Ioannoni and Y H Thong, 'The inhibitory effects of tetrandine on histamine release from rat mast cells.' Proc. Aust. Soc. Immunol. 213 (1988). Presented at National Australian Society for Immunology meeting in Canberra, February 1988.
21. 'B Ioannoni, A H Chalmers, W K Seow, J G McCormack and Y H Thong, 'Inhibition of the phosphatidylinositol secondary messenger system by tetrandine,' Proc. Aust. Soc. Immunol. 144 Immunology meeting in Canberra, February 1988.
22. W K Seow, A Ferrante, D B Goh, A H Chalmers, S Lo and Y H Thong, 'In vitro immunosuppressive properties of the plant alkaloid tetrandine,' Presented at the 6th Internation Meeting of Pharmacology in August 1987, Sydney.
23. D M Cowley, J M Brown, B C McWhinney and A H Chalmers, 'Hydroxycarboxylate malabsorption and calcium oxalate nephrolithiasis,' presented at the International Urolithiasis Meeting in July 1988, Vancouver, British Columbia, Canada. Urol. Res. 16, 179.
24. A H Chalmers, C Hare, G L Warren and I H Frazer (1989). *The Clinical Biochemists Reviews* 10, 103 (1989) Ectoenzymes as diagnostic markers of disease progression in HIV-infected

patients. Oral presentation at National AACB Meeting, Gold Coast, Queensland, October 1989.
25. A H Chalmers and F N Cornell (1989), *The Clinical Biochemists Reviews*, 10: 108 (1989) Affinity gel, rapid electrophoretic and fast performance liquid chromatographic methods for glycated haemoglobin measurements compared. Oral Presentation at National AACB Meeting, Gold Coast, Queensland October 1989.
26. R Seshadri, D Horsfall, A H Chalmers, V Settlur, K McCaul, 'The Prognostic significance of tumour cathepsin D levels ini primary breast cancer,' Presented at National Meeting of the Clinical Oncological Society of Australia meeting Sydney 1992.
27. B Ioannoni and A H Chalmers, 'Calcium Absorption in ilial brush border membrane vesicles: A Model for Calcium nephrolithiasis,' Presented at the Gordon Conference at Plymouth, New Hampshire, USA on Calcium Oxalate in June 1993.
28. A H Chalmers and B Ioannoni, 'Calcium uptake into ilial brush border membrane vesicles is similar in vivo in humans,' *The Clinical Biochemists Reviews* 14: 237 (1993). Presented at the XV International Congress of Clinical Chemistry, Melbourne, Australia, November 1993.
29. A H Chalmers, P J Wye and A J Bennett, 'Analysis of haemoglobin A2 and F in blood by fast performance liquid chromatography,' Presented at the National AACB meeting in October 1994, Adelaide. Clinical Biochemist Reviews 15: 134 (1994).
30. A H Chalmers and B Ioannoni, 'The role of citrate in causing calcium renal stone disease,' Oral Presentation at International Symposium on Human Stones, Trace Metals and Free Radicals, Udaipur, India, September 1995.
31. H Chalmers, J S Blake-Mortimer and A H Chalmers, '5'-Ectonucleotidase: a marker of oxidative damage to cells,' Invited Plenary Lecturer at the International Symposium on Human Stones, Trace Metals and Free Radicals, Udaipur, India, September 1995.
32. J S Blake-Mortimer, A H Winefield and A H Chalmers, 'The

relationship between 5' -ectonucleotidase and psychological stress. Psychoneuroimmunology' (in press), Presented at the International Scientific meeting of the Behavioural Immunology Group, Sydney, April 1995.

33. A H Chalmers, J Blake-Mortimer and A H Winefield, 'Evidence for superoxide anion mediated reduction of lymphocytic 5'-ectonucleotidase during stress,' Presented at the National Meeting of the AACB in Darwin, NT in August 1996. Clin Biochem Revs 17:79 (1996).

34. A H Chalmers, J Blake-Mortimer and A H Winefield, 'Lymphocytic 5'-ectonucleotidase, an indicator of superoxide anion activity in humans. Presented from 13–18 July 1997 at the Gordon Conference, Holderness College, Plymouth New Hampshire, USA.

35. J S Blake-Mortimer, A H Winefield and A H Chalmers, 'Lymphocytic 5'-ectonucleotidase: A marker of stress induced immune suppression. Research Perspectives in Psychoneuroimmunology V111. Abstracts of Psychoneuroimmunology Research Society meeting on Psychoneuroimmunomodulation,' vol 5. Presented in Bristol, England, March 1998.

36. A H Chalmers and M Kiley, 'Xanthochromia in CSF: Detection of sub-arachnoid haemorrhage,' Presented at the National Meeting of the AACB in Brisbane, Queensland in August 1998. Clin Biochem Rev 19: 72 1998. Oral presentation awarded first prize at the meeting.

37. A H Chalmers, J S Blake-Mortimer, A H Winefield, 'Lymphocytic 5'-ectonucleotidase, an indicator of superoxide anion activity in humans. Presented at Oxidative Pathways in Health and Disease,' Conference in Sydney, NSW in December 1999.

38. J R Hapuarachchi, A H Winefield, A H Chalmers, J S Blake-Mortimer, 'The impact of work-stress on health and wellbeing in university staff: changes in clinically relevant metabolites with psychological strain,' Presented at the International Congress of Behavioural Medicine, Helsinki, July 2002.

39. R Hapuarachchi, A H Winefield, A H Chalmers, J S Blake-

Mortimer, J Stough, N Gillespie, J Dua. 'The impact of work-stress on health and wellbeing in university staff-Changes in clinically relevant metabolites with psychological strain,' Presented at the 5th Industrial and Organisational Psychology Conference, Melbourne, June 2003.

www.ingramcontent.com/pod-product-compliance
Lightning Source LLC
Chambersburg PA
CBHW021428070526
44577CB00001B/106